Engineering Graphics Essentials
(A Text and Lecture Aid)

Kirstie Plantenberg
University of Detroit Mercy

ISBN: 1-58503-256-5

PUBLICATIONS

Schroff Development Corporation

www.schroff.com
www.schroff-europe.com

Copyright © 2005 Kirstie Platenberg

All rights reserved. No part of this book may be reproduced, stored in a retrieval system, or transcribed in any form or by any means – electronic, mechanical, photocopying, recording, or otherwise – without the prior written permission of Schroff Development Corporation.

PREFACE

Engineering Graphics Essentials is specifically designed to be used in a 1 or 2 credit introduction to engineering graphics course. It covers the main topics of engineering graphics, including tolerancing and fasteners, and gives engineering students a basic understanding of how to create and read engineering drawings. This text is also designed to encourage students to interact with the instructor during lecture. It has many examples that require student participation.

Supplements

Power Point lecture materials accompany the *Engineering Graphics Essentials* text. The presentations cover the entire book; however, they are segmented into basic and advanced topics. This allows instructors the flexibility to choose the material to be covered, based on the number of credits allocated to their course.

This book is dedicated to my family for their support and help.

Nassif, Summer and Lias Rayess
and
Phyllis Plantenberg

The following is a list of suggested chapters and topics that could be covered in a 1 credit engineering graphics class. It is possible for a 2 credit course to cover the entire book.

TABLE OF CONTENTS

NOTES:

ORTHOGRAPHIC PROJECTION

In Chapter 1 you will learn the importance of engineering graphics and how to create an orthographic projection. An orthographic projection describes the shape of an object. It is a two dimensional representation of a three dimensional object. Different line types are used to indicate visible, hidden and symmetry lines. By the end of this chapter, you will be able to create a technically correct orthographic projection using proper projection techniques.

1.1) INTRODUCTION TO ENGINEERING GRAPHICS

What is Engineering Graphics? Engineering graphics is a set of rules and guidelines that help you create an engineering drawing. *What is an Engineering Drawing?* An engineering drawing is a drawing or a set of drawings that communicates an idea, design, schematic, or model. Engineering drawings come in many forms. Each engineering field has its own type of engineering drawings. For example, electrical engineers draw circuit schematics and circuit board layouts. Civil engineers draw plans for bridges and road layouts. Mechanical engineers draw parts and assemblies that need to be manufactured. This book focuses on the latter. This is not to say that only students in a mechanical engineering curriculum will benefit from learning engineering graphics. It benefits everyone from the weekend carpenter who wants to draw plans for his/her new bookshelf to the electrical engineer who wants to analyze electrical component cooling using a CAE program. Engineering graphics teaches you how to visualize and see all sides of an object in your mind. Being able to visualize in your mind will help you in several aspects of critical thinking.

1.2) ORTHOGRAPHIC PROJECTION INTRODUCTION

An *orthographic projection* enables us to represent a 3-D object in 2-D (see Figure 1-1). An orthographic projection is a system of drawings that represent different sides of an object. These drawings are formed by projecting the edges of the object perpendicular to the desired planes of projection. Orthographic projections allow us to represent the shape of an object using 2 or more views. These views together with dimensions and notes are sufficient to manufacture the part.

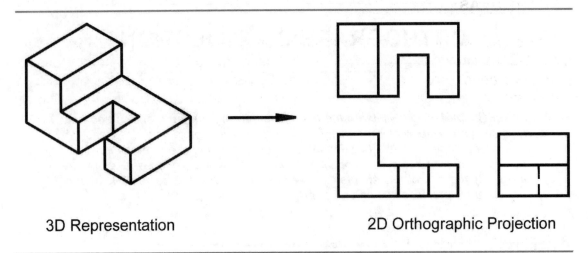

3D Representation   2D Orthographic Projection

Figure 1-1: Orthographic projection.

1.2.1) The Six Principle Views

The 6 principle views of an orthographic projection are shown in Figure 1-2. Each principle view is created by looking at the object in the directions indicated in Figure 1-2 and drawing what is seen as well as what is hidden from view.

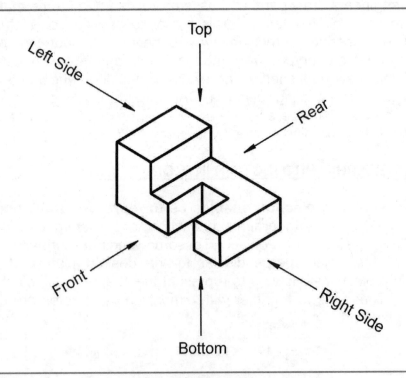

Figure 1-2: The six principle views.

1.3) **THE GLASS BOX METHOD**

To obtain an orthographic projection, an object is placed in an imaginary glass box as shown in Figure 1-3. The sides of the glass box represent the six principle planes. Images of the object are projected onto the sides of the box to create the six principle views. The box is then unfolded to lie flat, showing all views in a 2-D plane. Figure 1-4 shows the glass box being unfolded to create the orthographic projection of the object.

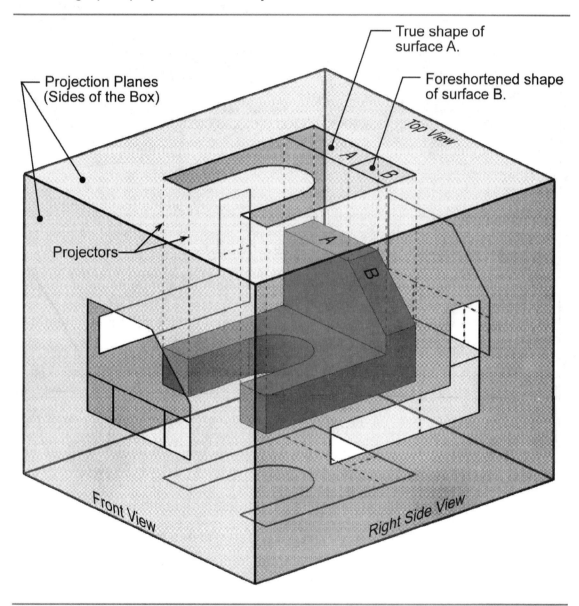

Figure 1-3: Object in a glass box.

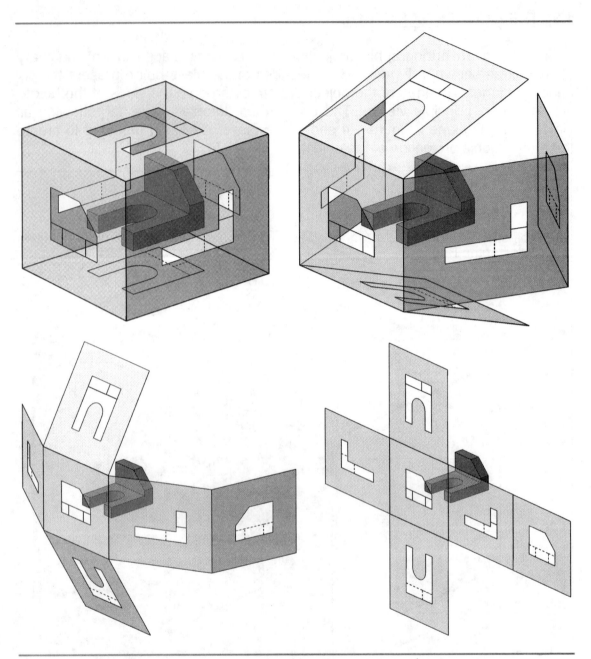

Figure 1-4: Glass box being unfolded.

Instructor Led Exercise 1-1: Principle views

Label the five remaining principle views with the appropriate view name.

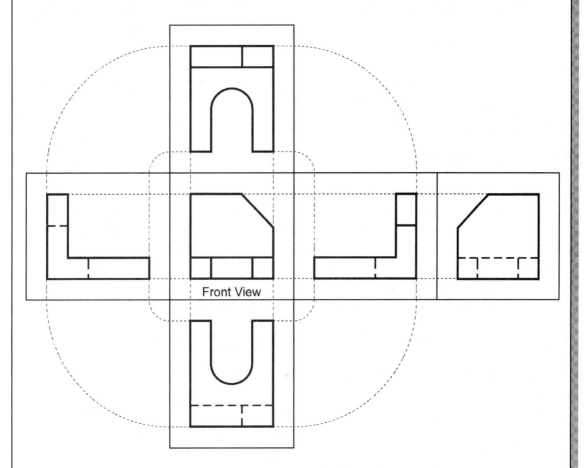

Front View

What are the differences between the *Right Side* and *Left Side* views?

What are the differences between the *Top* and *Bottom*, and *Front* and *Rear* views?

Which view(s) have the least number of hidden or dashed lines?

1.4) <u>THE STANDARD VIEWS</u>

When constructing an orthographic projection, we need to include enough views to completely describe the true shape of the part. The more complex a part, the more views are needed to describe it completely. Most objects require three views to completely describe them. **The standard views used in an orthographic projection are the *front, top,* and *right side* views.** The other views (bottom, rear, left side) are omitted since they usually do not add any new information. It is not always necessary to use the three standard views. Some objects can be completely described in one or two views. For example, a sphere only requires one view, and a block only requires two views.

1.4.1) <u>The Front View</u>

The *front view* shows the most features or characteristics of the object. It usually contains the least number of hidden lines. The exception to this rule is when the object has a predefined or generally accepted front view. All other views are based on the orientation chosen for the front view. The top, front, and bottom views are all aligned vertically and share the same width dimension. The left side, front, right side, and rear views are all aligned horizontally and share the same height dimension (see the figure shown in Exercise 1-1).

1.5) <u>LINE TYPES USED IN AN ORTHOGRAPHIC PROJECTION</u>

***Line type* and *line weight* provide valuable information to the print reader.** For example, the type and weight of a line can answer the following questions: Is the feature visible or hidden from view? Is the line part of the object or part of a dimension? Is the line indicating symmetry? There are four commonly used line types: continuous, hidden, center and phantom. The standard recommends using, no less than, two line widths. Important lines should be twice as thick and as the less important thin lines. Common thicknesses are 0.6 mm for important lines and 0.3 mm for the less important lines. However, to further distinguish line importance, it is recommended to use four different thicknesses or weights: thin, medium, thick, and very thick. The actual line thickness should be chosen such that there is a visible difference between the line weights; however, they should not be too thick or thin making it difficult to read the print. The thickness of the lines should be adjusted according to the size and complexity of the part. The following is a list of common line types and widths used in an orthographic projection.

1. <u>Visible lines:</u> Visible lines represent visible edges and boundaries. The line type is **continuous** and the line weight is **thick** (0.5 - 0.6 mm).

2. <u>Hidden lines:</u> Hidden lines represent edges and boundaries that cannot be seen. The line type is **dashed** and the line weight is **medium thick** (0.35 - 0.45 mm).

3. Center lines: Center lines represent axes of symmetry and are important for interpreting cylindrical shapes. Crossed center lines should be drawn at the centers of circles. They are also used to indicate circle of centers and paths of motion. The line type is **long dash – short dash** and the line weight is **thin** (0.3 mm).

4. Phantom lines: Phantom lines are used to indicate imaginary features. For example, they are used to indicate the alternate positions of moving parts, and adjacent positions of related parts. The line type is **long dash – short dash – short dash** and the line weight is usually **thin** (0.3 mm).

5. Dimension and Extension lines: Dimension and extension lines are used to show the size of an object. In general, a dimension line is placed between two extension lines and is terminated by arrowheads, which indicates the direction and extent of the dimension. The line type is **continuous** and the line weight is **thin** (0.3 mm).

6. Cutting plane lines: Cutting plane lines are used to show where an imaginary cut has been made through the object in order to view interior features. The line type is **phantom** and the line weight is **very thick** (0.6 to 0.8 mm). Arrows are placed at both ends of the cutting plane line to indicate the direction of sight.

7. Section lines: Section lines are used to show areas that have been cut by the cutting plane. Section lines are grouped in parallel line patterns and usually drawn at a 45° angle. The line type is usually **continuous** and the line weight is **thin** (0.3 mm).

8. Break lines: Break lines are used to show imaginary breaks in objects. A break line is usually made up of a series of connecting arcs. The line type is **continuous** and the line weight is usually **thick** (0.5 – 0.6 mm).

<u>Instructor Led Exercise 1-2: Line types</u>

Using the line type definitions given on the previous page, draw each line type listed.

- Visible Line

- Hidden Line

- Center Line

- Phantom Line

- Dimension and Extension Lines

- Cutting Plane Line

- Section Lines

- Break Line

Instructor Led Exercise 1-3: Line use in an orthographic projection

Fill the following dotted orthographic projection with the appropriate line types.

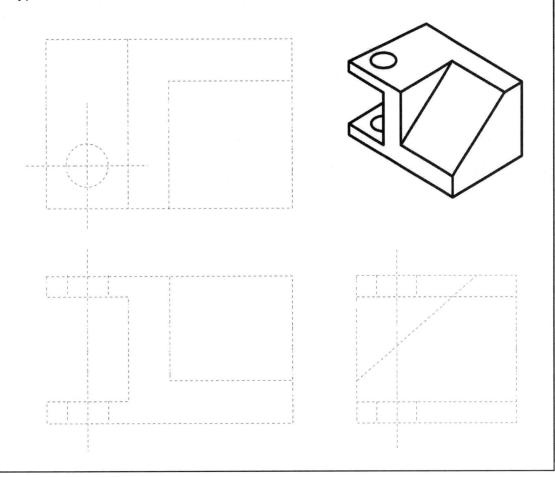

1.6) RULES FOR LINE CREATION AND USE

The rules and guide lines for line creation should be followed in order to create lines that are effective in communicating the drawing information. However, due to computer automation, some of the rules may be hard to follow.

1.6.1) Hidden Lines

Hidden lines represent edges and boundaries that cannot be seen.

Rule 1. The length of the hidden line dashes may vary slightly as the size of the drawing changes. For example, a very small part may require smaller dashes in order for the hidden line to be recognized.

Rule 2. Hidden lines should always begin and end with a dash, except when the hidden line begins or ends at a parallel visible or hidden line (see Figure 1-5).

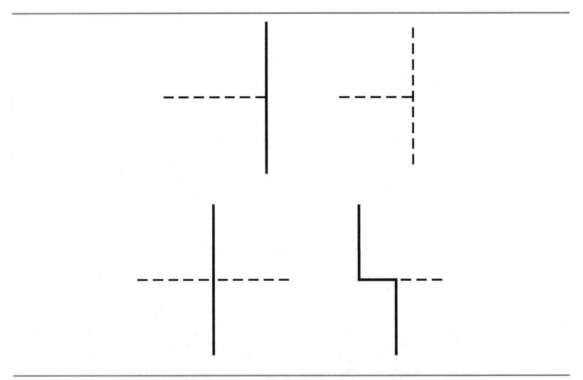

Figure 1-5: Drawing hidden lines.

Rule 3. Dashes should join at corners (see Figure 1-6).

Figure 1-6: Hidden lines at corner.

1.6.2) Center Lines

Center lines represent axes of symmetry and are important for interpreting cylindrical shapes. They are also used to indicate circle of centers, as shown in Figure 1-7, and paths of motion.

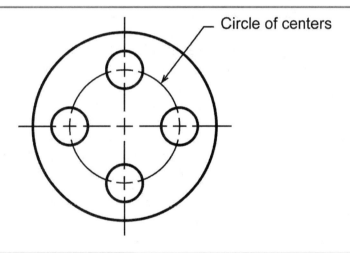
Circle of centers

Figure 1-7: Circle of centers.

Rule 1. Center lines should start and end with long dashes (see Figure 1-8).

Rule 2. Center lines should intersect by crossing either the long dashes or the short dashes (see Figure 1-8).

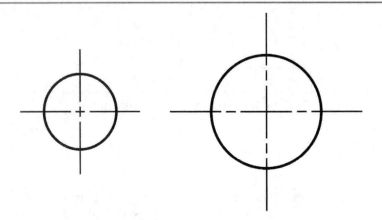

Figure 1-8: Crossing center lines.

Rule 3. Center lines should extend a short distance beyond the object or feature. They should not terminate at other lines of the drawing (see Figure 1-9).

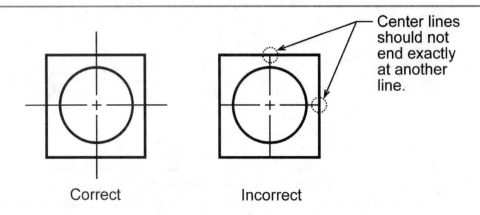

Figure 1-9: Terminating center lines.

Rule 4. Center lines may be connected within a single view to show that two or more features lie in the same plane as shown in Figure 1-10. However, they should not extend through the space between views.

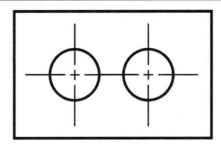

Figure 1-10: Connecting center lines.

1.6.3) Phantom Lines

Phantom lines should start and end with a long dash. Phantom lines are used to indicate alternate positions of moving parts. They may also be used to indicate adjacent positions of related parts and repeated detail. They are also used to show fillets and rounds in the view that does not show the radius. In this case, the phantom lines are used to show a change in surface direction. Examples of phantom line use are shown in Figures 1-11 and 1-12.

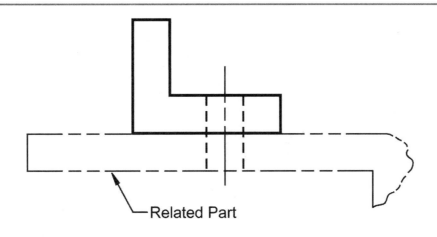

Related Part

Figure 1-11: Related part.

Figure 1-12: Repeated detail.

1.6.4) <u>Break Lines</u>

Break lines are used to show imaginary breaks in an object. For example, when drawing a long rod, it may be broken and drawn at a shorter length as shown in Figure 1-13.

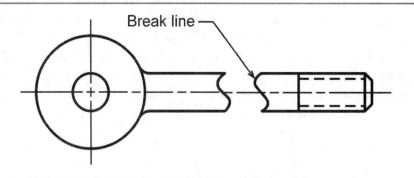

Figure 1-13: Using break lines.

There are two types of break lines. A break line may be a series of connecting arcs, as shown in Figure 1-13, or a straight line with a jog in the middle as shown in Figure 1-14. If the distance to traverse is short the series of connecting arcs is used. This series of arcs is the same width as the visible lines on the drawing. If the distance is long the thin straight line with a jog is used.

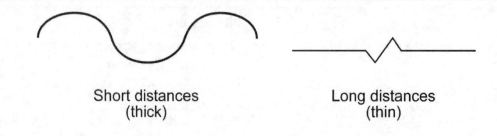

Figure 1-14: Types of break lines.

1.6.5) Line Type Precedence

Some lines are considered more important than other lines. **If two lines occur in the same place, the line that is considered to be the least important is omitted.** Lines in order of precedence/importance are as follows:

1. Cutting plane line
2. Visible line
3. Hidden line
4. Center line

1.7) CREATING AN ORTHOGRAPHIC PROJECTION

The steps presented in this section are meant to help you create a technically correct orthographic projection using the 3rd angle projection standard. To understand and visually see how views are created using the 3rd angle projection standard, put your right hand on a table palm up. You are looking at the front view of your hand. Now rotate your hand so that your thumb points up and your little finger is touching the table. This is the right side view of your hand. Put your hand back in the front view position. Now rotate your hand so that your finger tips are pointing up and your wrist is touching the table. This is the top view of your hand.

The following steps will take you through the creation of an orthographic projection. Once you become experienced and proficient at creating orthographic projections, you will develop short cuts and may not need to follow the steps exactly as written. These steps are visually illustrated in Figure 1-15.

1. **Choose a front view.** This is the view that shows the most about the object.
2. **Decide how many views are needed** to completely describe the object. If you are unable to determine which views will be needed, draw the standard views (front, top and right side).
3. **Draw the visible features of the front view.**
4. **Draw projectors off of the front view** horizontally and vertically in order to create the boundaries for the top and right side views.
5. **Draw the top view.** Use the vertical projectors to fill in the visible and hidden features.
6. **Project from the top view back to the front view.** Use the vertical projectors to fill in any missing visible or hidden features in the front view.
7. **Draw a 45° projector** off of the upper right corner of the box that encloses the front view.
8. **From the top view, draw projectors over to the 45° line and down** in order to create the boundaries of the right side view.
9. **Draw the right side view.**
10. **Project back to the top and front view** from the right side view as needed.
11. **Draw center lines where necessary.**

Following the aforementioned steps will insure that the orthographic projection is done correctly. That is, it will insure that:

√ The front and top views are vertically aligned.
√ The front and right side views are horizontally aligned.
√ Every point or feature in one view is aligned on a projector in any adjacent view (front and top, or front and right side).
√ The distance between any two points of the same feature in the related views (top and right side) are equal.

Figure 1-15 identifies the *adjacent* and *related* views. Adjacent views are two adjoining views aligned by projectors. Related views are views that are adjacent to the same view.

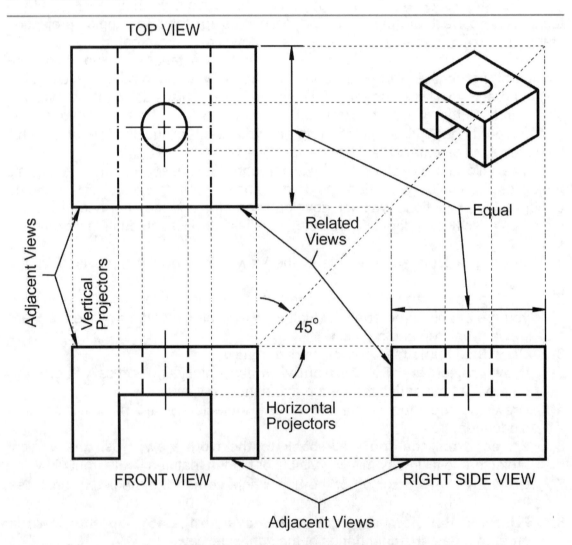

Figure 1-15: Creating an orthographic projection

1.7.1) Projection Symbol

In the United States, we use 3rd angle projection to create an orthographic projection. This is the method of creating orthographic projections that is described in this chapter. In some parts of Europe and elsewhere 1st angle projection is used. To inform the print reader what projection method was used, the projection symbol should be placed in the bottom right hand corner of the drawing. If the drawing uses metric units, the text "SI" is placed in front of the projection symbol. The projection symbols are shown in Figure 1-16. Figure 1-17 shows the projection symbol's proportions.

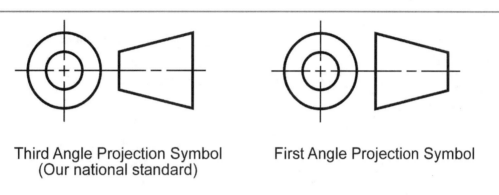

Third Angle Projection Symbol
(Our national standard)

First Angle Projection Symbol

Figure 1-16: First and third angle projection symbols.

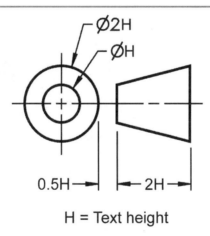

H = Text height

Figure 1-17: Projection symbol proportions.

NOTES:

In Class Student Exercise 1-4: Missing lines 1

Name: _____   Date: _____

Fill in the missing lines in the front, right side, and top views. **Hint:** The front view has one missing visible line. The right side view has one missing visible line and two missing hidden lines. The top view has five missing visible lines and two missing hidden lines.

TOP

FRONT   RIGHT SIDE

NOTES:

In Class Student Exercise 1-5: Missing lines 2

Name: _____ Date: _____

Fill in the missing lines in the top, front, and right side views. **Hint:** The top view has one missing visible line. The front view has four missing visible lines and four missing center lines. The right side view has two missing hidden lines and one missing center line.

TOP

FRONT   RIGHT SIDE

NOTES:

In Class Student Exercise 1-6: Drawing a orthographic projection 1

Name: _____   Date: _____

Shade in the surfaces that will appear in the front, top, and right side views. Estimating the distances, draw the front, top, and right side views. Identify the surfaces with the appropriate letter in the orthographic projection.

TOP

FRONT   RIGHT SIDE

NOTES:

In Class Student Exercise 1-7: Drawing orthographic projection 2

Name: _____   Date: _____

Identify the best choice for the front view. Estimating the distances, draw the front, top, and right side views.

1.8) <u>AUXILIARY VIEWS</u>

Auxiliary views are used to show the true shape of features that are not parallel to any of the principle planes of projection. Auxiliary views are aligned with the angled features from which they are projected. Partial auxiliary views are often used to shown only a particular feature that is not described by true projection in the principle views. Figure 1-18 shows the use of auxiliary views.

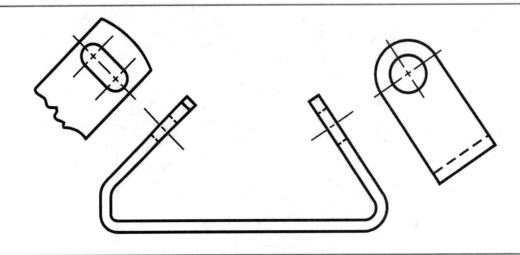

Figure 1-18: Auxiliary views.

In Class Student Exercise 1-8: Auxiliary view

Name: _____ Date: _____

Draw the auxiliary view for this object.

NOTES:

ORTHOGRAPHIC PROJECTION REVIEW QUESTIONS

Name: _____ Date: _____

Answer the following questions.

Q1-1) Is an orthographic projection a 2-D or 3-D representation of an object?

Q1-2) Center lines should end at the boundary of an object. (True, False)

Q1-3) List two situations where a phantom line would be used?

Q1-4) What is the thickest line on a non-sectioned orthographic projection?

Q1-5) What are the standard views used in an orthographic projection?

Q1-6) Are the top and front views aligned vertically or horizontally?

Q1-7) Are the front and right side views aligned vertically or horizontally?

Q1-8) Are projection or construction lines usually shown on a drawing?

Q1-9) Which view generally contains the least number of hidden lines?

Q1-10) Why do we use different line types and line weights?

Q1-11) What line type is used to indicate symmetry?

Q1-12) Rank the following line types in order of precedence, 1 being the most important.
   a. Center line
   b. Visible
   c. Hidden

Q1-13)   The United States uses (1st , 3rd) angle projection.

Q1-14)   In what situation would an auxiliary view be used?

Q1-15)   In what situation would a detail view be used?

ORTHOGRAPHIC PROJECTION PROBLEMS

Name: _____ Date: _____

P 1-1) Sketch the front, top and right side views of the following object. Use the grid provided.

<u>NOTES:</u>

Name: _____ Date: _____

P 1-2) Sketch the front, top and right side views of the following object. Use the grid provided.

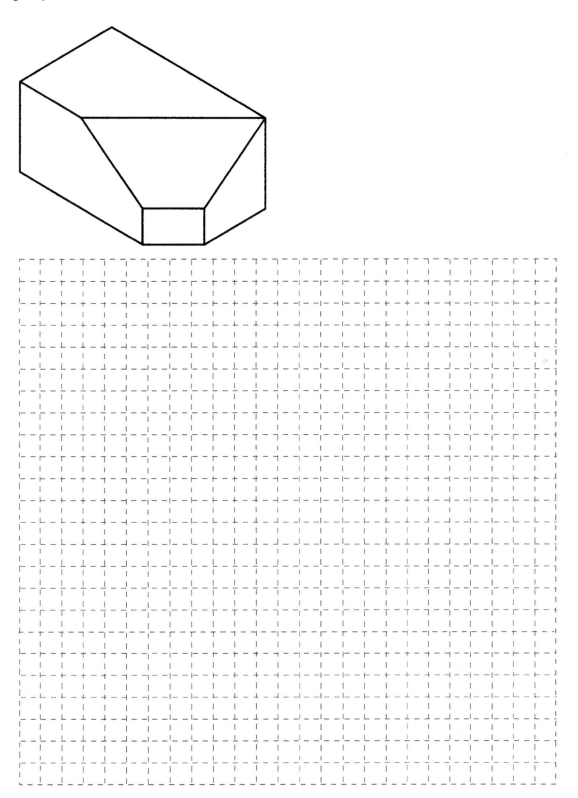

NOTES:

Name: _____ Date: _____

P 1-3) Sketch the front, top and right side views of the following object. Use the grid provided.

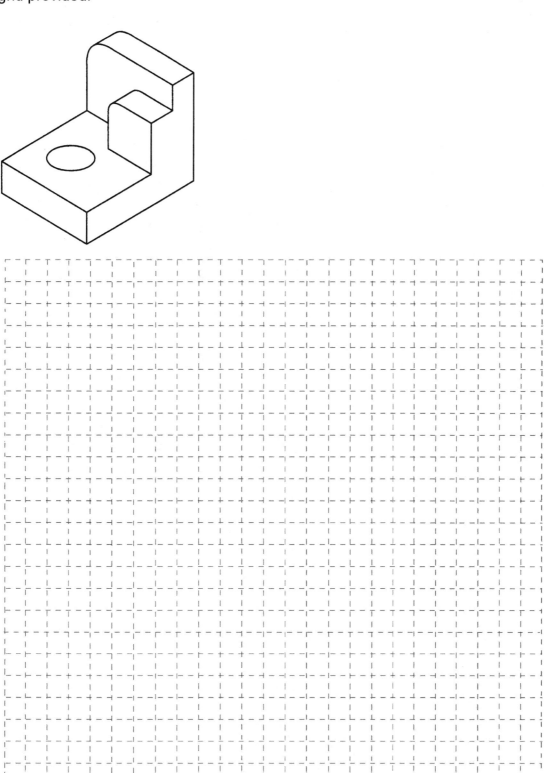

NOTES:

Name: _____ Date: _____

P 1-4) Sketch the front, top and right side views of the following object. Use the grid provided.

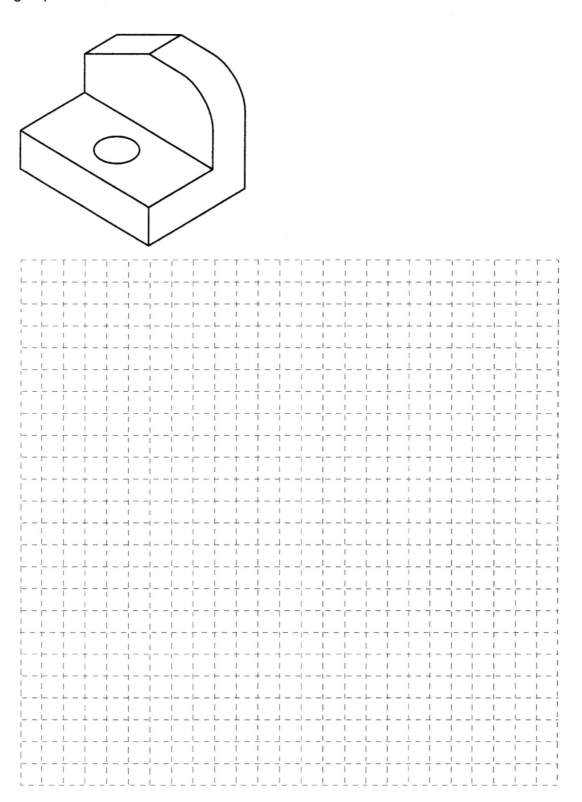

<u>NOTES:</u>

Name: _____ Date: _____

P 1-5) Sketch the front, top and right side views of the following object. Use the grid provided.

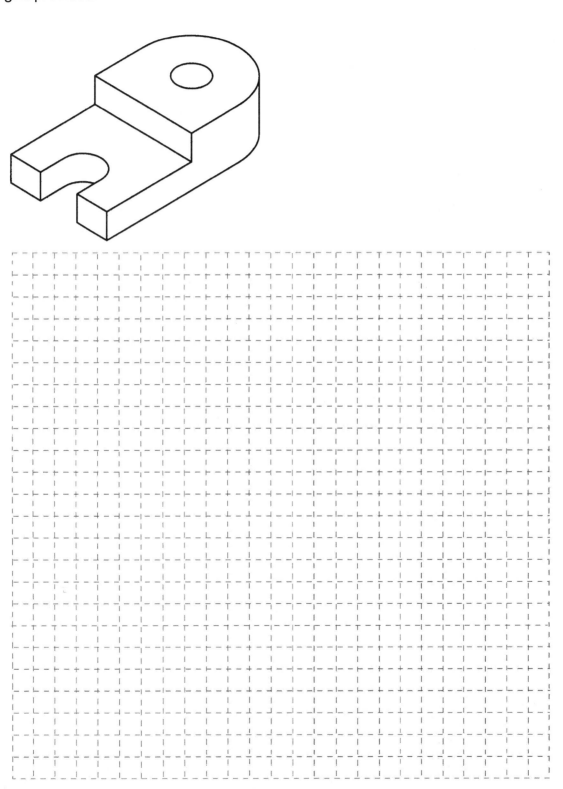

NOTES:

Name: _____ Date: _____

P 1-6) Sketch the front, top and right side views of the following object. Use the grid provided.

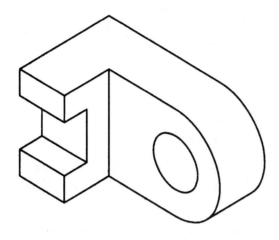

NOTES:

Name: _____ Date: _____

P 1-7) Sketch the front, top and right side views of the following object. Use the grid provided.

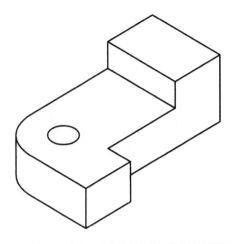

NOTES:

Name: _____ Date: _____

P 1-8) Sketch the front, top and right side views of the following object. Use the grid provided.

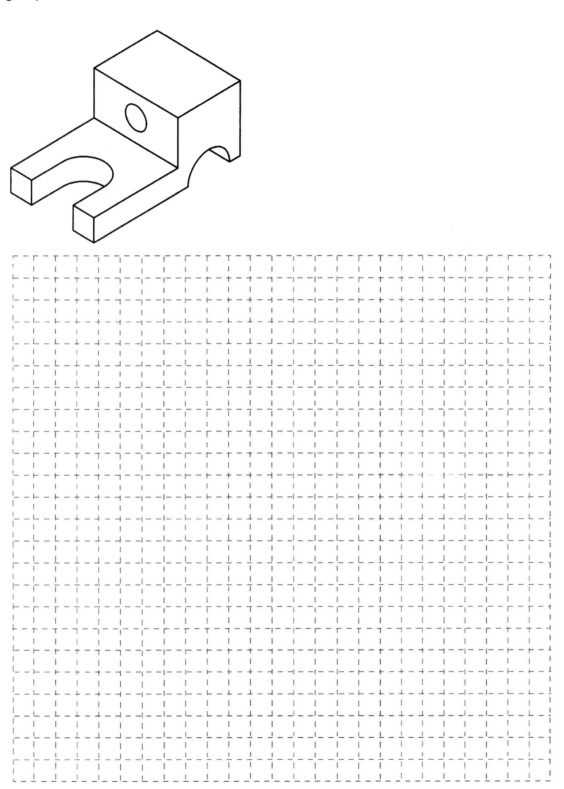

NOTES:

Name: _____ Date: _____

P 1-9) Sketch the front, top and right side views of the following object. Use the grid provided.

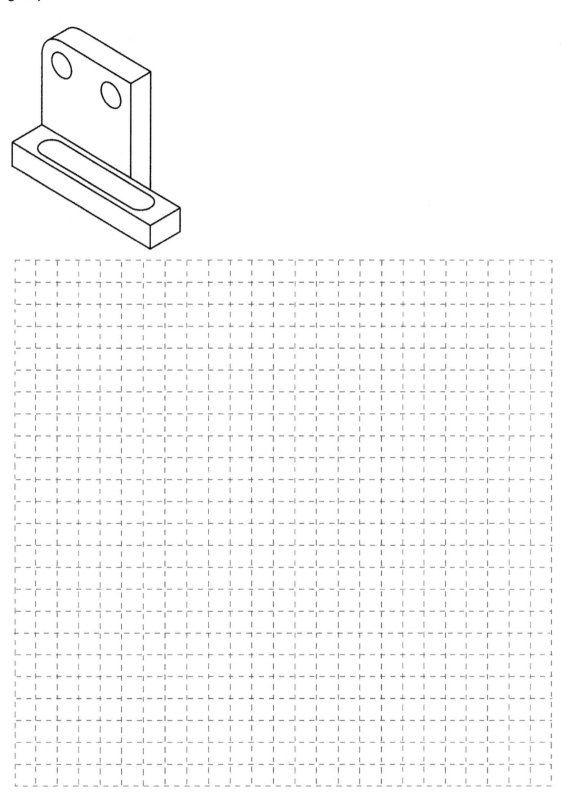

NOTES:

Name: _____ Date: _____

P 1-10) Sketch the front, top and right side views of the following object. Use the grid provided.

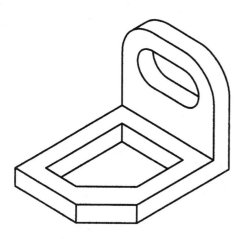

NOTES:

Name: _____ Date: _____

P 1-11) Sketch the front, top and right side views of the following object. Use the grid provided.

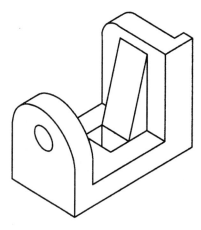

NOTES:

Name: _____ Date: _____

P 1-12) Sketch the front, top and right side views of the following object. Use the grid provided.

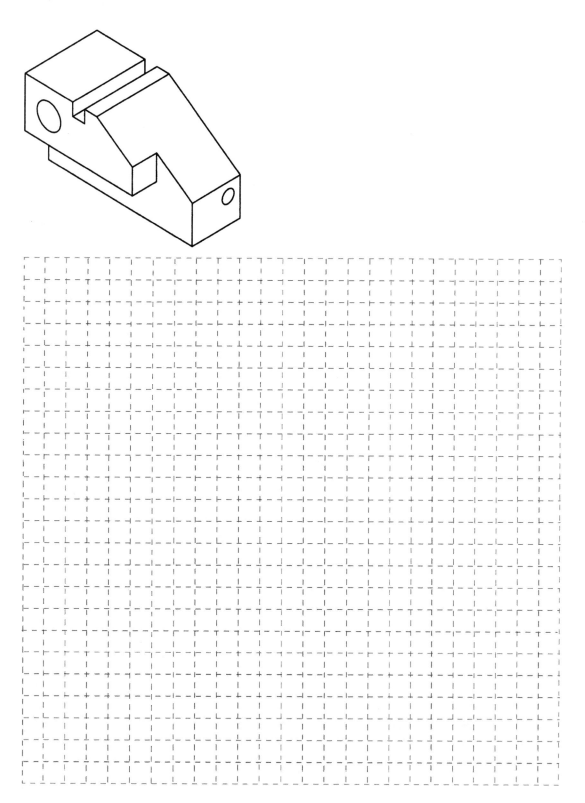

NOTES:

P1-13) Use your class CAD package to create an orthographic projection of the following object. Draw the three standard views.

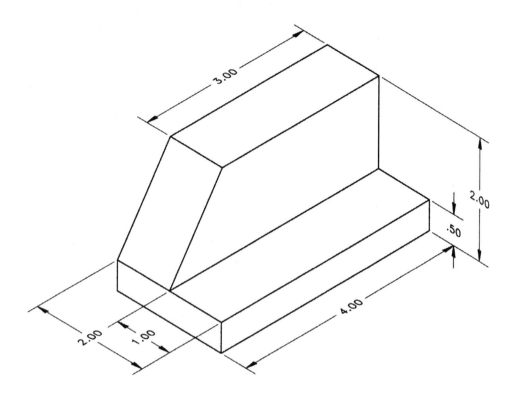

P1-14) Use your class CAD package to create an orthographic projection of the following object. Draw the three standard views.

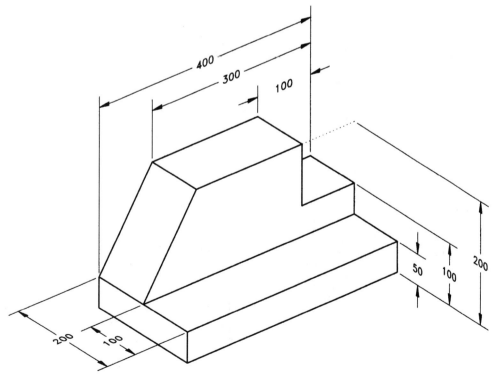

P1-15) Use your class CAD package to create an orthographic projection of the following object. Draw the three standard views.

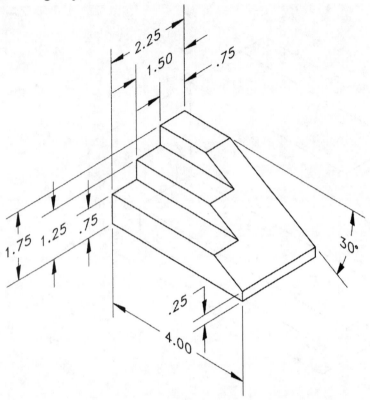

P2-16) Use your class CAD package to create an orthographic projection of the following object. Draw the three standard views.

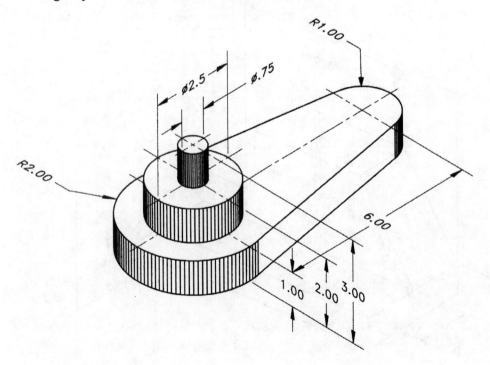

P1-17) Use your class CAD package to create an orthographic projection of the following object. Draw the three standard views.

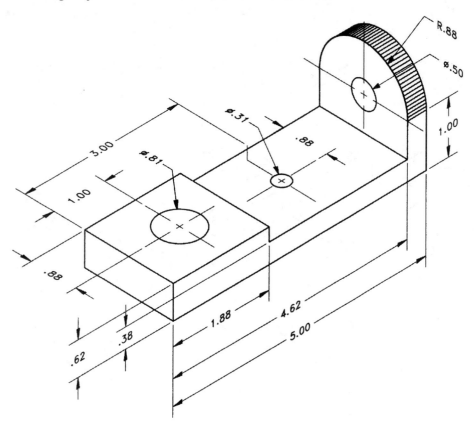

P1-18) Use your class CAD package to create an orthographic projection of the following object. Draw the three standard views.

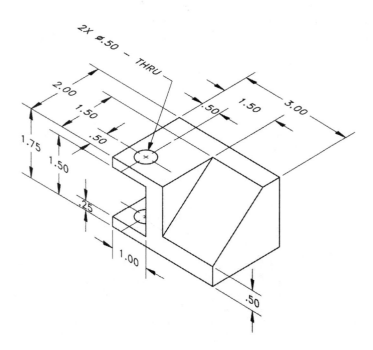

P1-19) Use your class CAD package to create an orthographic projection of the following object. Draw the three standard views.

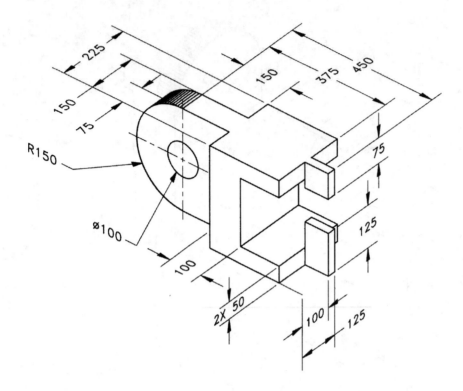

P1-20) Use your class CAD package to create an orthographic projection of the following object. Draw the three standard views.

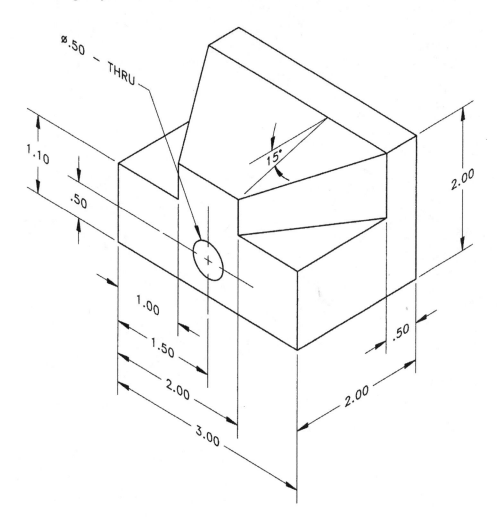

P1-21) Use your class CAD package to create an orthographic projection of the following object. Draw the three standard views.

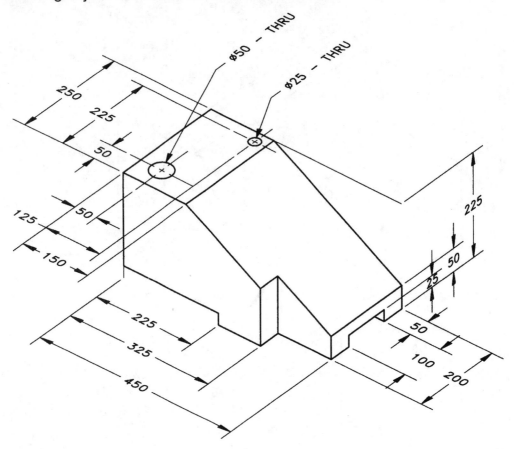

DIMENSIONING

In Chapter 2 you will learn how to dimension an orthographic projection using proper dimensioning techniques. This may seem like a simple task; however, dimensioning a part is not as easy as inserting the sizes used to draw the part. Dimensions affect how a part is manufactured. A small change in how an object is dimensioned may produce a part that will not pass inspection. The type and placement of the dimensions and the dimension text is highly controlled by ASME standards (American Society of Mechanical Engineers). By the end of this chapter, you will be able to dimension a moderately complex part using proper dimensioning techniques. CAUTION! Dimensioning complex/production parts require the knowledge of GD&T (Geometric Dimensioning & Tolerancing). However, GD&T is too complex a topic for an introductory class in engineering graphics.

2.1) DETAILED DRAWINGS

In addition to the shape description of an object given by the orthographic projection, an engineering drawing must also give a complete size description using dimensions. This enables the object to be manufactured. **An orthographic projection, complete with all the dimensions and specifications needed to manufacture the object is called a *detailed drawing*.** Figure 2-1 shows an example of a detailed drawing.

Dimensioning a part correctly entails conformance to many rules. It is very tempting to dimension an object using the measurements needed to draw the part. But, these are not necessarily the dimensions required to manufacture it. Generally accepted dimensioning standards should be used when dimensioning any object. Basically, the dimensions should be given in a clear and concise manner and should include everything needed to produce and inspect the part exactly as intended by the designer. There should be no need to measure the size of a feature directly from the drawing.

The dimensioning standards presented in this chapter are in accordance with the ASME Y14.5M-1994 standard. Other common sense practices will also be presented.

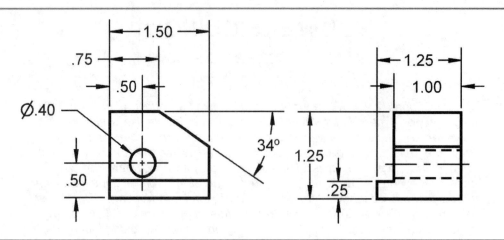

Figure 2-1: Detailed drawing

2.2) LEARNING TO DIMENSION

Proper dimensioning techniques require the knowledge of the following three areas.

1) <u>Dimension Appearance and Techniques:</u> Dimensions use special lines, arrows, symbols and text. In Section 2.3 (Dimension Appearance and Techniques) we will learn:

    a) The lines used in dimensioning.
    b) Types of dimensions.
    c) Dimension symbols.
    d) Dimension spacing and readability.
    e) Dimension placement.

2) <u>Dimensioning and Locating Features:</u> Different types of features require unique methods of dimensioning.

3) <u>Dimension Choice:</u> Your choice of dimensions will directly influence the method used to manufacture a part. Learning the following topics will guide you when choosing your dimension units, decimal places and the dimension's starting point:

    a) Units and decimal places.
    b) Locating features using datums.
    c) Dimension accuracy and error build up.

2.3) **DIMENSION APPEARANCE AND TECHNIQUES**

2.3.1) **Lines Used in Dimensioning**

Dimensioning requires the use of *dimension*, *extension* and *leader lines*. **All lines used in dimensioning are drawn thin so that they will not be confused with visible lines.** Thin lines should be drawn at approximately 0.3 mm or 0.016 inch.

- Dimension line: A dimension line is a thin solid line terminated by arrowheads, which indicates the direction and extent of a dimension. A number is placed near the mid point to specify the part's size.

- Extension line: An extension line is a thin solid line that extends from a point on the drawing to which the dimension refers. The dimension line meets the extension lines at right angles, except in special cases. **There should be a visible gap between the extension line and the object.** Long extension lines should be avoided.

Figure 2-2 illustrates the different features of a dimension.

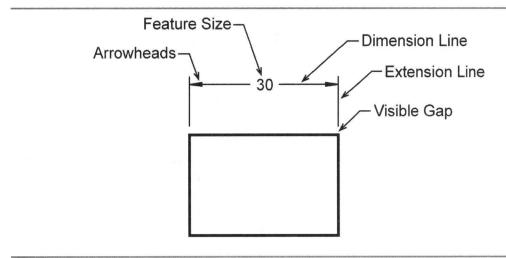

Figure 2-2: Features of a dimension.

- Leader line: A leader line is a straight inclined thin solid line that is usually terminated by an arrowhead. It is used to direct a dimension, note, symbol, item number, or part number to the intended feature on a drawing. The leader is not vertical or horizontal, except for a short horizontal portion extending to the first or last letter of the note. The horizontal part should not underline the note and may be omitted entirely.

The leader may be terminated:
   a) with an arrow, if it ends on the outline of an object.
   b) with a dot ($\emptyset$1.5 mm, minimum), if it ends within the outline of an object.
   c) without an arrowhead or dot, if it ends within the outline of an object.

When creating leader lines, the following should be avoided:
   a) Crossing leaders.
   b) Long leaders.
   c) Leaders that are parallel to adjacent dimension, extension or section lines.
   d) Small angles between the leader and the terminating surface.

Figure 2-3 illustrates different leader line configurations.

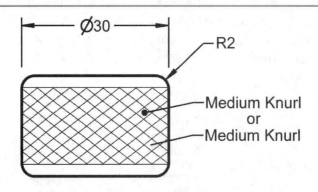

Figure 2-3: Leader line configurations.

- Arrowheads: The length and width ratio of an arrowhead should be 3 to 1 and the width should be proportional to the line thickness. A single style of arrowhead should be used throughout the drawing. Arrowheads are drawn between the extension lines if possible. If space is limited, they may be drawn on the outside. Figure 2-4 shows the most common arrowhead configurations.

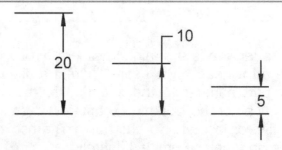

Figure 2-4: Arrowhead and feature size placement.

2.3.2) Types of Dimensions

Dimensions are given in the form of *linear distances*, *angles*, and *notes*.

- Linear distances: A linear dimension is used to give the distance between two points. They are usually arranged horizontally or vertically, but may also be aligned with a particular feature of the part.

- Angles: An angular dimension is used to give the angle between two surfaces or features of a part.

- Notes: Notes are used to dimension diameters, radii, chamfers, threads, and other features that can not be dimensioned by the other two methods.

Instructor Led Exercise 2-1: Dimension types

In Figure 2-1, count the different types of dimensions.

- How many linear horizontal dimensions are there?

- How many linear vertical dimensions are there?

- How many angular dimensions are there?

- How many leader line notes are there?

2.3.3) Lettering

Lettering should be legible, easy to read, and uniform throughout the drawing. Upper case letters should be used for all lettering unless a lower case is required. **The minimum lettering height is 0.12 in (3 mm).**

2.3.4) Dimensioning Symbols

Dimensioning symbols replace text and are used to minimize language barriers. Many companies produce parts all over the world. A print made in the U.S.A. may have to be read in several different countries. The goal of using dimensioning symbols it to eliminate the need for language translation. Table 2-1 shows some commonly used dimensioning symbols. These symbols will be used and explained throughout the chapter. Size and proportion of these symbols are given in Appendix C.

Term	Symbol	Term	Symbol
Diameter	⌀	Depth / Deep	▽
Spherical diameter	S⌀	Dimension not to scale	<u>10</u>
Radius	R	Square (Shape)	□
Spherical radius	SR	Arc length	$\widehat{5}$
Reference dimension	(8)	Conical Taper	▷
Counterbore / Spotface	⊔	Slope	◁
Countersink	∨	Symmetry	≑
Number of places	4X		

Table 2-1: Dimensioning symbols.

2.3.5) Dimension Spacing and Readability

Dimensions should be easy to read and minimize the possibility for conflicting interpretations. Dimensions should be given clearly and in an organized fashion. They should not be crowded or hard to read. The following is a list of rules that control dimension spacing and readability:

a) The spacing of dimension lines should be uniform throughout the drawing. The space between the first dimension line and the part should be at least 10 mm; the space between subsequent dimension should be at least 6 mm. However, the above spacing is only intended as a guide.

b) Do not dimension inside an object or have the dimension line touch the object unless clearness is gained.

c) Dimension text should be horizontal which means that it is read from the bottom of the drawing.

d) Dimension text should not cross dimension, extension or visible lines.

<u>Instructor Led Exercise 2-2: Spacing and readability 1</u>

Consider the incorrectly dimensioned object shown.   There are six dimensioning mistakes. List them and then dimension the object correctly.

1)   4)
2)   5)
3)   6)

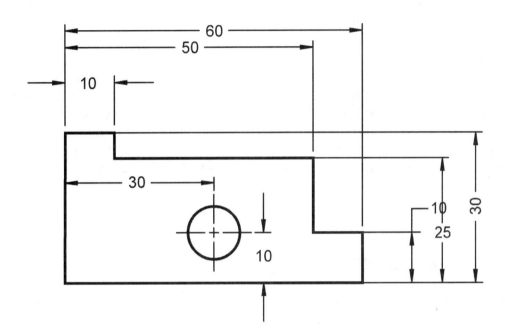

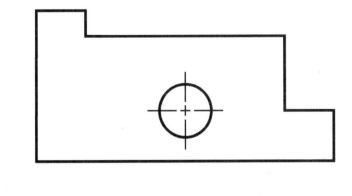

e) Dimension lines should not cross extension lines or other dimension lines. To avoid this, shorter dimensions should be placed before longer ones. Extension lines can cross other extension lines or visible lines. However, this should be minimized. Where extension lines cross other lines, the extension lines are not broken. If an extension line crosses an arrowhead or is near an arrowhead, a break in the extension line is permitted.

f) Extension lines and centerlines should not connect between views.

g) Leader lines should be straight, not curved, and point to the center of the arc or circle at an angle between 30°-60°.

Try Exercise 2-3.

h) Dimensions should not be duplicated or the same information given in two different ways. The use of reference (duplicated) dimensions should be minimized. Duplicate dimensions may cause needless trouble. If a change is made to one dimension, the reference dimension may be overlooked causing confusion. If a reference dimension is used, the size value is placed within parentheses (e.g. (10)).

Try Exercise 2-4.

2.3.6) Dimension Placement

Dimensions should be placed in such a way as to enhance the communication of your design. The following are rules that govern the logical and practical arrangement of dimensions to insure maximum legibility:

a) Dimensions should be grouped whenever possible.

b) Dimensions should be placed between views, unless clearness is promoted by placing some outside.

c) Dimensions should be attached to the view where the shape is shown best.

d) Do not dimension hidden lines.

Try Exercise 2-5.

<u>Instructor Led Exercise 2-3: Spacing and readability 2</u>

Consider the incorrectly dimensioned object shown. There are four dimensioning mistakes. List them and then dimension the object correctly.

1)   3)
2)   4)

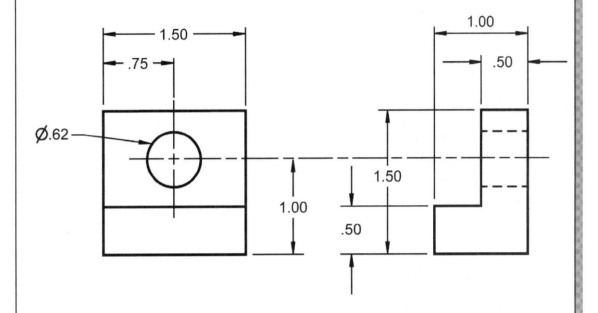

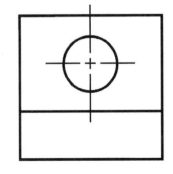

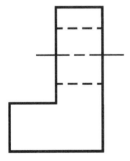

Instructor Led Exercise 2-4: Duplicate dimensions

Find the duplicate dimensions and cross out the ones that you feel should
be omitted.

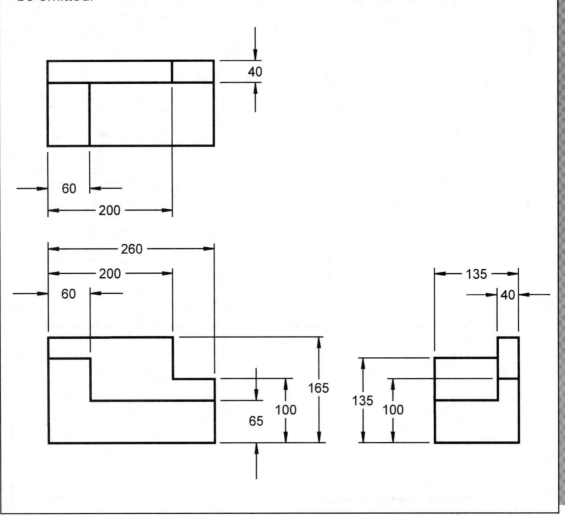

Instructor Led Exercise 2-5: Dimension placement

Consider the incorrectly dimensioned object shown. There are six dimensioning mistakes. List them and then dimension the object correctly.

1)   4)
2)   5)
3)   6)

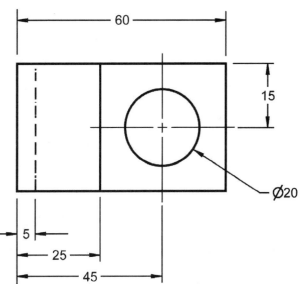

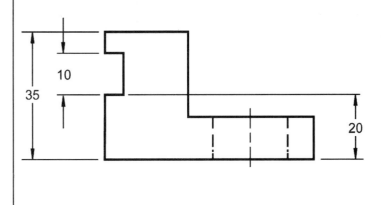

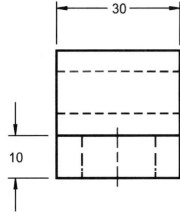

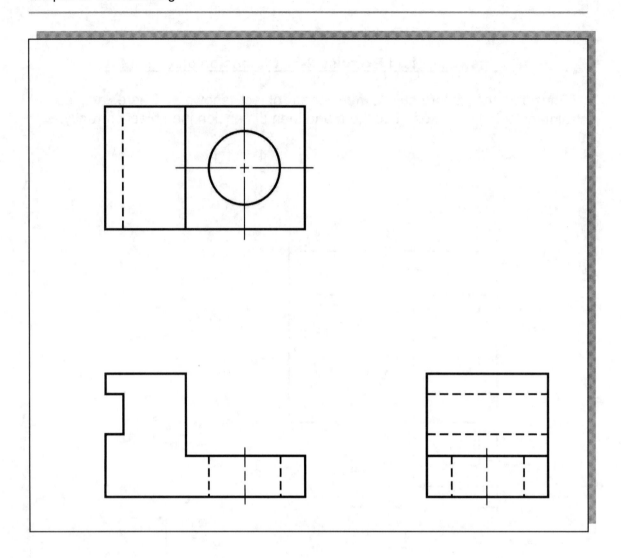

2.4) <u>DIMENSIONING AND LOCATING SIMPLE FEATURES</u>

The following section illustrates the standard ways of dimensioning different basics features that occur often on a part.

a) A circle is dimensioned by its diameter and an arc by its radius using a leader line and a note. A diameter dimension is preceded by the symbol "∅", and a radial dimension is preceded by the symbol "R". On older drawings you may see the abbreviation "DIA" placed after a diameter dimension and the abbreviation "R" following a radial dimension. Figure 2-5 illustrates the diameter and radius dimensions.

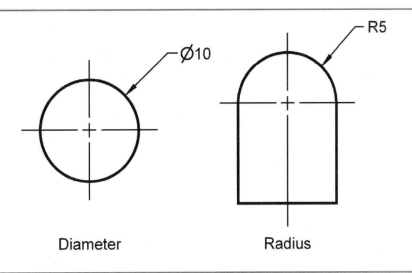

Figure 2-5: Diameter and radius dimensions

Instructor Led Exercise 2-6: Circular and rectangular views

Below is shown the front and top view of a part. Consider the hole and cylinder features of the part when answering the following questions.
- Which view is considered the circular view and which is considered the rectangular view?
- Looking at just the top view, can you tell the difference between the hole and the cylinder?

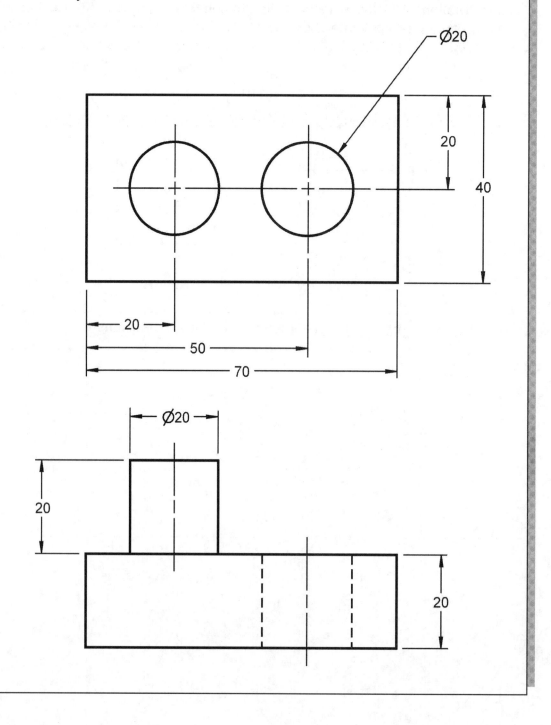

b) Holes are dimensioned by giving their diameter and location in the circular view (see Exercise 2-6).

c) A cylinder is dimensioned by giving its diameter and length in the rectangular view, and is located in the circular view. By giving the diameter of a cylinder in the rectangular view, it is less likely to be confused with a hole (see Exercise 2-6).

d) Repetitive features or dimensions may be specified by using the symbol "X" along with the number of times the feature is repeated. There is no space between the number of times the feature is repeated and the "X" symbol, however, there is a space between the symbol "X" and the dimension.

Instructor Led Exercise 2-7: Dimensioning and locating features

Dimension the following object.

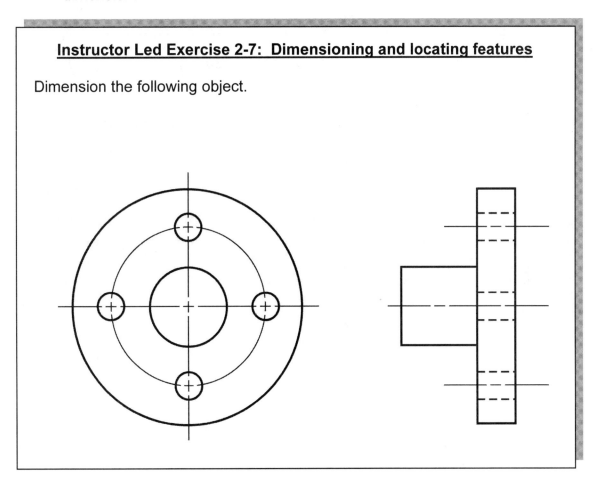

2.5) DIMENSIONING AND LOCATING ADVANCED FEATURES

The following section will illustrate the standard ways of dimensioning different features that occur often on a part.

a) If a dimension is given to the center of a radius, a small cross is drawn at the center. Where the center location of the radius is unimportant, the drawing must clearly show that the arc location is controlled by other dimensioned features such as tangent surfaces. Figure 2-6 shows several different types of radius dimensions.

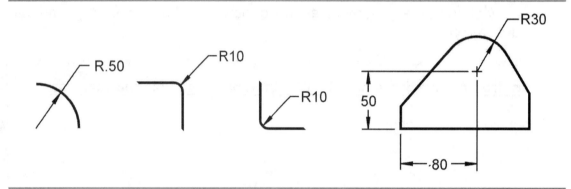

Figure 2-6: Dimensioning radial features.

b) A complete sphere is dimensioned by its diameter and an incomplete sphere by its radius. A spherical diameter is indicated by using the symbol "S∅" and a spherical radius by the symbol "SR". Figure 2-7 illustrates the spherical diameter and spherical radius dimensions.

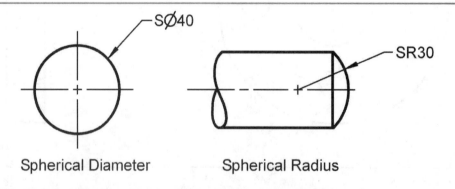

Figure 2-7: Dimensioning spherical features.

c) The depth of a blind hole may be specified in a note and is the depth of the full diameter from the surface of the object. Figure 2-8 illustrates how to dimension a blind hole (i.e. a hole that does not pass completely through the object).

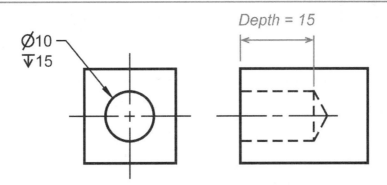

Figure 2-8: Dimensioning a blind hole.

d) If a hole goes completely through the feature and it is not clearly shown on the drawing, the abbreviation "THRU" follows the dimension.

e) If a part is symmetric, it is only necessary to dimension to one side of the center line of symmetry. The center line of symmetry is indicated by using the symbol "⟰". On older drawings you might see the symbol "₵" used instead. Figure 2-9 illustrates the use of the symmetry symbol.

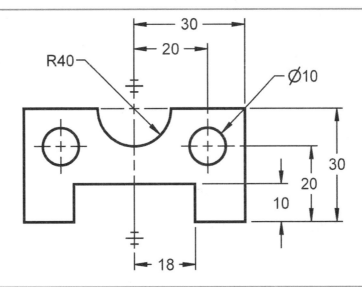

Figure 2-9: Center line of symmetry.

f) Counterbored holes are specified by giving the diameter of the drill (and depth if appropriate), the diameter (∅) of the counterbore (⊔), and the depth (▽) of the counterbore in a note as shown in Figure 2-10. If the thickness of the material below the counterbore is significant, this thickness rather than the counterbore depth is given.

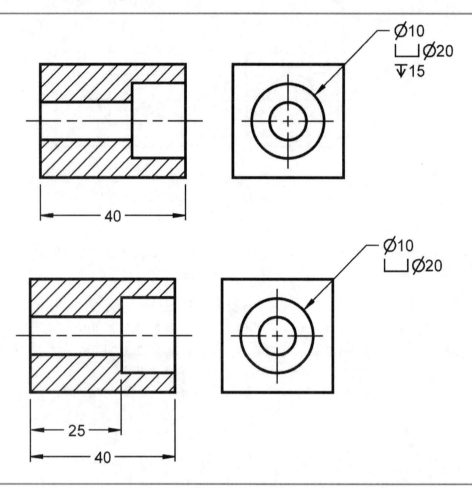

Figure 2-10: Counterbored holes.

Application Question 2-1

What is the purpose of a counterbored hole?

g) Spotfaced holes are similar to counterbored holes. The difference is that the machining operation occurs on a curved surface. Therefore, the depth of the counterbore drill can not be given in the note. It must be specified in the rectangular view as shown in Figure 2-11.

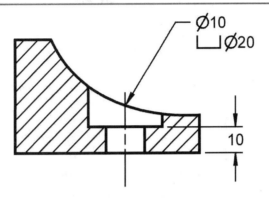

Figure 2-11: Spotfaced holes.

h) Countersunk holes are specified by giving the diameter of the drill (and depth if appropriate), the diameter of the countersink ($\vee$), and the angle of the countersink in a note as shown in Figure 2-12.

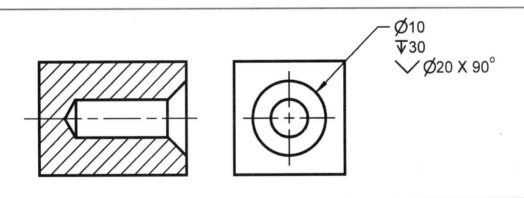

Figure 2-12: Countersunk holes.

Application Question 2-2

What is the purpose of a countersunk hole?

---

i) Chamfers are dimensioned by a linear dimension and an angle, or by two linear dimensions. A note may be used to specify 45 degree chamfers because the linear value applies in either direction (see Figure 2-13). Notice that there is a space between the 'X' symbol and the linear dimension. The space is inserted so that it is not confused with a repeated feature.

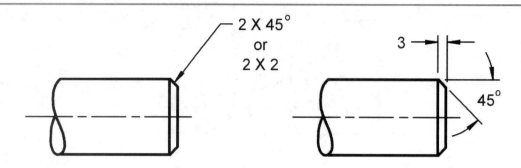

Figure 2-13: Chamfers.

Application Question 2-3

What is the purpose of a chamfer?

2.5.1) Drawing Notes

Drawing notes give additional information that is used to complement conventional dimension. Drawing notes provide information that clarify the manufacturing requirements for the part. They cover information such as treatments and finishes among other manufacturing processes. A note may also be used to give blanket dimensions, such as the size of all rounds and fillets on a casting or a blanket tolerance. Notes may apply to the entire drawing or to a specific area. A general note applies to the entire drawing. A local note is positioned near and points to the specified area to which it applies. The note area is identified with the heading "NOTE:".

Instructor Led Exercise 2-8: Advanced features

Consider the incorrectly dimensioned object shown. There are seven dimensioning mistakes. List them and then dimension the object correctly.

1)   5)
2)   6)
3)   7)
4)

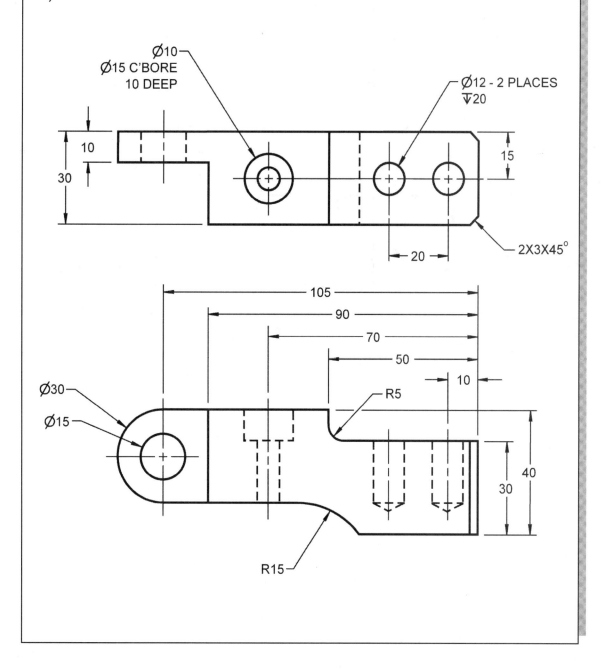

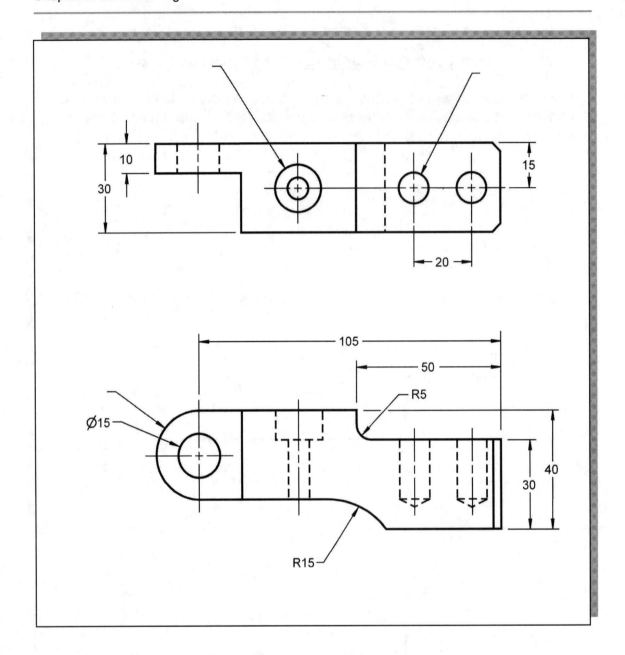

2.6) <u>DIMENSION CHOICE</u>

Dimension placement and dimension text influences the manufacturing process used to make the part. However, your **choice of dimensions should depend on the function and the mating relationship of the part**, and then on manufacturing. Even though dimensions influence how the part is made, the manufacturing process should not be specifically stated on the drawing.

2.6.1) <u>Units and Decimal Places</u>

a) Decimal dimensions should be used for all machining dimensions. Sometimes you may encounter a drawing that specifies standard drills, broaches, and the like by size. For drill sizes that are given by number or letter, a decimal size should also be given.

b) Metric dimensions are given in 'mm' and to 0 or 1 decimal place (e.g. 10, 10.2). When the dimension is less than a millimeter, a zero should proceed the decimal point (e.g. 0.5).

c) English dimensions are given in 'inches' and to 2 decimal places (e.g. 1.25). A zero is not shown before the decimal point for values less than one inch (e.g. .75).

d) Metric 3rd angle drawings are designated by the SI projection symbol shown in Figure 2-14.

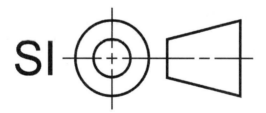

Figure 2-14: SI projection symbol.

2.6.2) Locating Features Using Datums

Consider three mutually perpendicular datum planes as shown in Figure 2-15. These planes are imaginary and theoretically exact. Now, consider a part that touches all three datum planes. The surfaces of the part that touch the datum planes are called datum features. **Most of the time, features on a part are located with respect to a datum feature.** In some cases, it is necessary to locate a feature with respect to another feature that is not the datum feature.

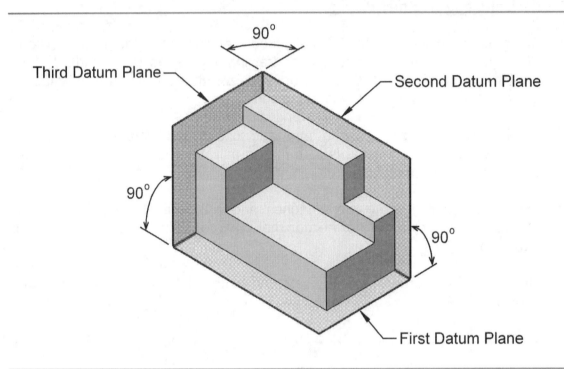

Figure 2-15: Datums and datum features.

Datum feature selection is based on the function of the part. When selecting datum features, think of the part as a component of an assembly. Functionally important surfaces and features should be selected as datum features. For example, to ensure proper assembly, mating surfaces should be used as datum features. A datum feature should be big enough to permit its use in manufacturing the part. If the function of the part is not known, take all possible measures to determine its function before dimensioning the part. In the process of learning proper dimensioning techniques, it may be necessary to make an educated guess as to the function of the part. Figure 2-16 shows a dimensioned part. Notice how all the dimensions originate from the datum features.

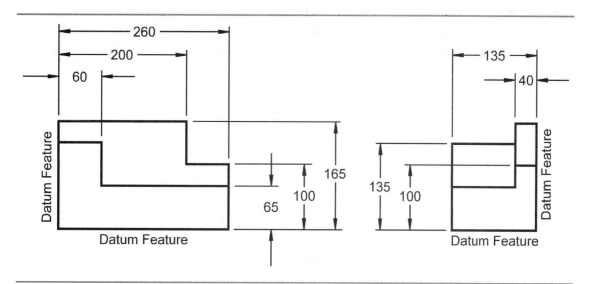

Figure 2-16: Dimensioning using datum features.

a) Datum dimensioning is preferred over continuous dimensioning (see Figure 2-17). Features should be located with respect to datum features.

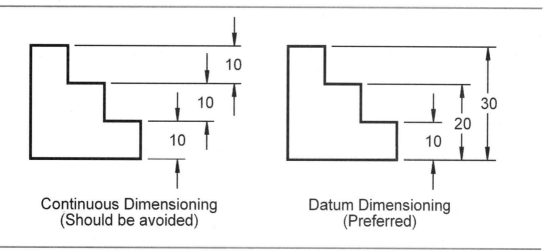

Continuous Dimensioning
(Should be avoided)

Datum Dimensioning
(Preferred)

Figure 2-17: Continuous versus datum dimensioning.

b) Dimensions should be given between points or surfaces that have a functional relation to each other (slots, mating hole patterns, etc...). Figure 2-18 shows a part that has two holes that are designed to mate up with two holes on another part.

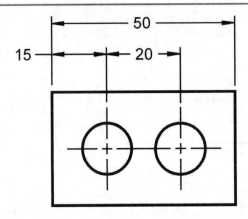

Figure 2-18: Dimensioning functionally important features.

Application Question 2-4

Referring to the part shown in Figure 2-18, why is the distance between the two holes functionally important?

2.6.3) Dimension Accuracy

There is no such thing as an "exact" measurement. Every dimension has an implied or stated tolerance associated with it. A tolerance is the amount a dimension is allowed to vary.

Instructor Led Exercise 2-9: Dimension accuracy

Consider the figure shown below.

- Which dimensions have implied tolerances and which have stated tolerances?

- Does the arrow indicate an increasing or decreasing accuracy?

- Write down the range in which the dimension values are allowed to vary.

  (a)

  (b)

  (c)

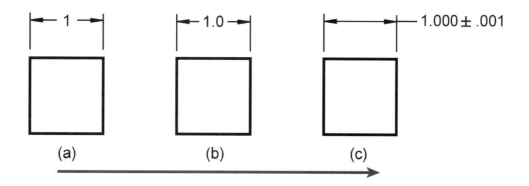

2.6.4) <u>Rounding off</u>

The more accurate the dimension the more expensive it is to manufacture. To cut costs it is necessary to round off fractional dimensions. If, for example, we are rounding off to the second decimal place and the third decimal place number is less than 5, we truncate after the second decimal place. If the number in the third decimal place is greater than 5, we round up and increase the second decimal place number by 1. If the number is exactly 5, whether or not we round up depends on if the second decimal place number is odd or even. If it is odd, we round up and if it is even, it is kept the same.

<u>Instructor Led Exercise 2-10: Rounding off</u>

Round off the following fractions to two decimal places according to the rules stated above.

(5/16) .3125 →   (1/8) .125 →

(5/32) .1562 →   (3/8) .375 →

2.6.5) <u>Cumulative Tolerances (Error Buildup)</u>

Figure 2-19 shows two different styles of dimensioning. One is called *Continuous Dimensioning*, the other *Datum Dimensioning*. Continuous dimensioning has the disadvantage of accumulating error. **It is preferable to use datum dimensioning to reduce error buildup.**

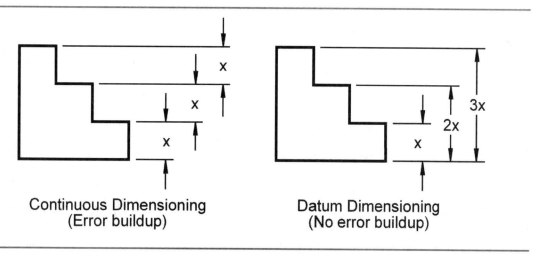

Continuous Dimensioning
(Error buildup)

Datum Dimensioning
(No error buildup)

Figure 2-19: Error buildup.

Consider the part shown in Figure 2-19. It is dimensioned using both continuous and datum dimensioning. The implied tolerance of all the dimensions is on the first decimal place. If we look at the continuous dimensioning case, the actual dimensions are $x.e$, where $0.e$ is the error associated with each dimension. Adding up the individual dimensions, we get an overall dimension of $3x + 3*(0.e)$. The overall dimension for the datum dimensioning case is $3x + 0.e$. As this example shows, continuous dimensioning accumulates error.

Another advantage of using datum dimensioning is the fact that many manufacturing machines are programmed using a datum or origin. Therefore, it makes it easier for the machinist to program the machine if datum dimensioning is used.

Instructor Led Exercise 2-11: Dimension choice

Consider the incorrectly dimensioned object shown. There are five dimensioning mistakes. List them and then dimension the object correctly.

1)   4)

2)   5)

3)

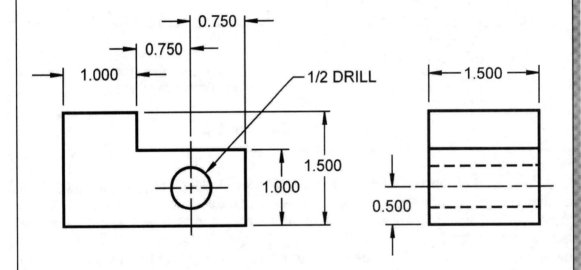

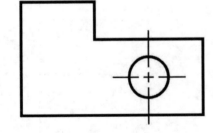

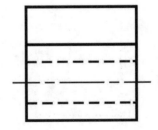

In Class Student Exercise 2-12: Dimensioning 1

Name: _____ Date: _____

Dimension the following object using proper dimensioning techniques. Did we need to draw the right side view?

This way is up

NOTES:

In Class Student Exercise 2-13: Dimensioning 2

Name: _____ Date: _____

Dimension the following object using proper dimensioning techniques.

This way is up

NOTES:

In Class Student Exercise 2-14: Dimensioning 3

Name: _____ Date: _____

Dimension the following object using proper dimensioning techniques.

This way is up

NOTES:

In Class Student Exercise 2-15: Dimensioning 4

Name: _____   Date: _____

Dimension the following object using proper dimensioning techniques.

NOTES:

DIMENSIONING REVIEW QUESTIONS

Name: _____ Date: _____

Answer the following questions.

Q2-1) What is the difference between an orthographic projection and a detailed drawing?

Q2-2) Explain the difference between an extension line, a dimension line, and a leader line.

Q2-3) Why are dimension and extension lines thin?

Q2-4) Can a leader line be horizontal or vertical?

Q2-5) What are the three types of dimensions?

Q2-6) Dimensioning hidden lines under some circumstances is allowed. (True, False)

Q2-7) What is meant when referring to the rectangular view of a hole?

Q2-8) A circular hole is dimensioned by its (radius, diameter) in the (circular, rectangular) view, and the center is located in the (circular, rectangular) view.

Q2-9) A cylinder should be located in the (circular, rectangular) view and its diameter given in the (circular, rectangular) view.

Q2-10) Give two reasons why datum dimensioning is preferred over continuous dimensioning.

Q2-11) What units of measurement are most commonly used on prints? List one for a metric drawing and one for an English drawing.

Q2-12) How is two thousandths of an inch expressed on a blue print?

Q2-13) In general, inch dimensions are given to (1, 2, 3, 4) decimal places and millimeter dimensions are given to (0, 1, 2, 3) decimal places.

Q2-14) Explain the significance of the following symbol.

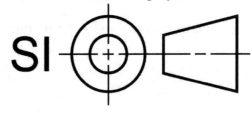

Q2-15) How are reference dimensions shown on a print?

Q2-16) What symbol is used to indicate repeated features? Is there any situation in which this symbol is used in another context?

Q2-17) What symbol is used to indicate the depth of a blind hole?

Q2-18) What symbols are used to indicate counterbored and countersunk holes?

Name: _____ Date: _____

Q2-19) What is the purpose of a drawing note?

Q2-20) What is a datum feature?

Q2-21) When should an auxiliary view be used?

NOTES:

DIMENSIONING PROBLEMS

Name: _____   Date: _____

P2-1) The following object is dimensioned incorrectly. Identify the incorrect dimensions and list all mistakes associated with them. Then, dimension the object correctly using proper dimensioning techniques. There are five mistakes.

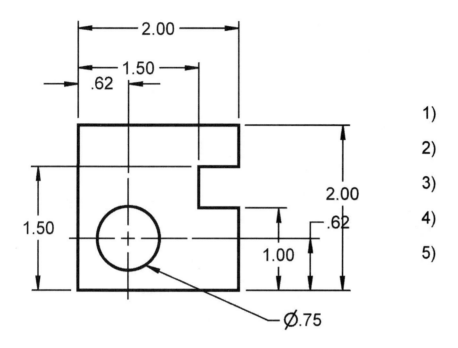

1)

2)

3)

4)

5)

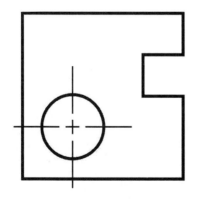

NOTES:

Name: _____   Date: _____

P2-2) The following object is dimensioned incorrectly. Identify the incorrect dimensions and list all mistakes associated with them. Then, dimension the object correctly using proper dimensioning techniques. There are four mistakes.

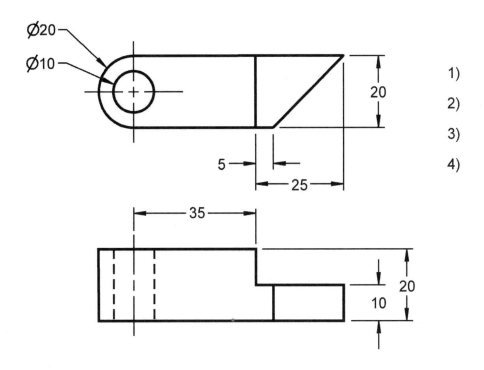

1)

2)

3)

4)

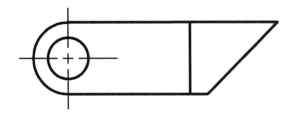

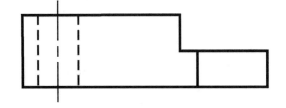

NOTES:

Name: _____ Date: _____

P2-3) The following object is dimensioned incorrectly. Identify the incorrect dimensions and list all mistakes associated with them. Then, dimension the object correctly using proper dimensioning techniques. There are six mistakes.

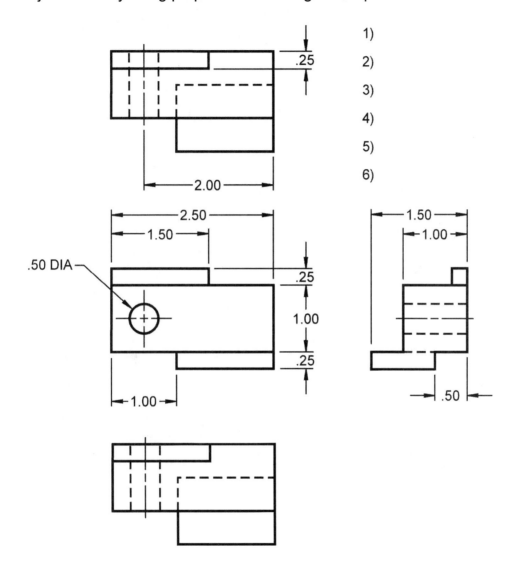

1)

2)

3)

4)

5)

6)

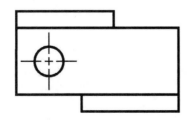

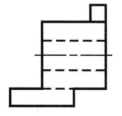

<u>NOTES:</u>

Name: _____ Date: _____

P2-4) The following object is dimensioned incorrectly. Identify the incorrect dimensions and list all mistakes associated with them. Then, dimension the object correctly using proper dimensioning techniques. There are four mistakes.

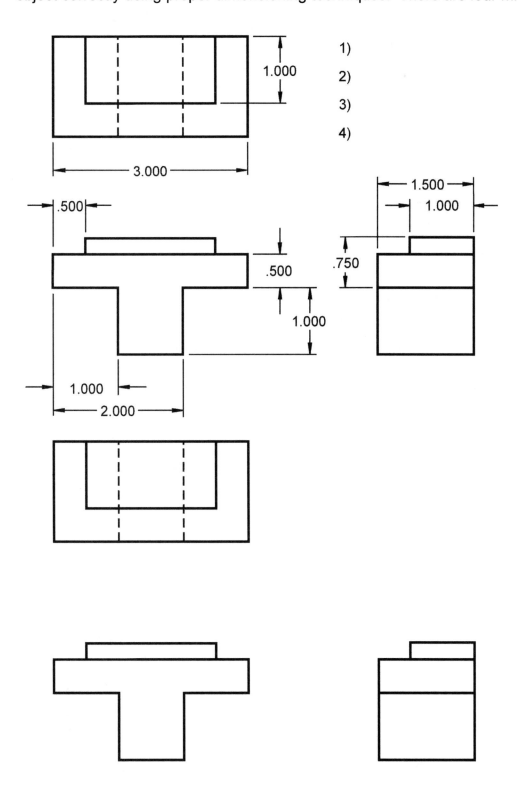

1)

2)

3)

4)

NOTES:

Name: _____   Date: _____

P2-5) The following object is dimensioned incorrectly. Identify the incorrect dimensions and list all mistakes associated with them. Then, dimension the object correctly using proper dimensioning techniques. There are five mistakes.

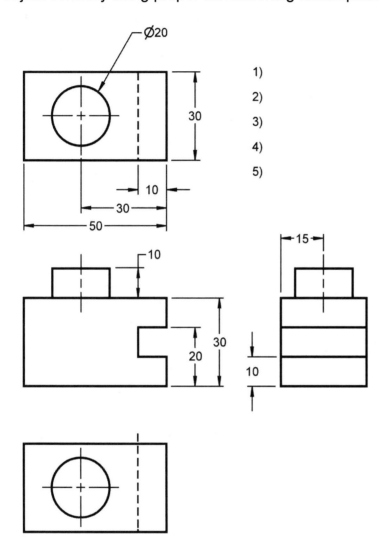

1)

2)

3)

4)

5)

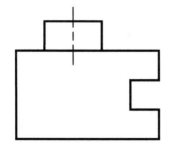

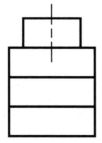

<u>NOTES:</u>

Name: _____   Date: _____

P2-6)   Completely dimension the objects shown (by hand) using proper dimensioning techniques. Wherever a numerical dimension value is required, place an 'x'. Use dimensioning symbols where necessary.

a)

b)

c)

d)

e)

f)

NOTES:

Name: _____   Date: _____

P2-7) Completely dimension the objects shown (by hand) using proper dimensioning techniques. Wherever a numerical dimension value is required, place an 'x'. Use dimensioning symbols where necessary.

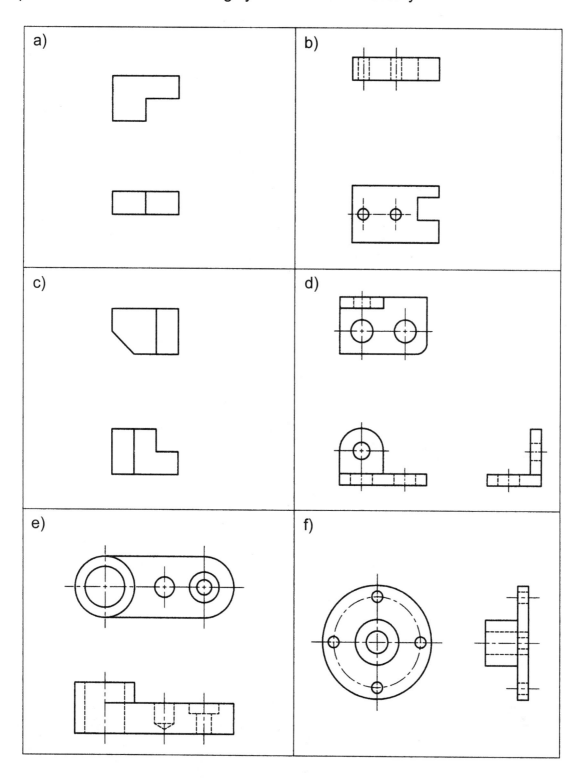

<u>NOTES:</u>

P2-8) Using a CAD package, draw the necessary views and completely dimension the objects given. Do not base your 2-D dimension placement on the 3-D dimensions shown. Use proper dimensioning techniques to dimension your object.

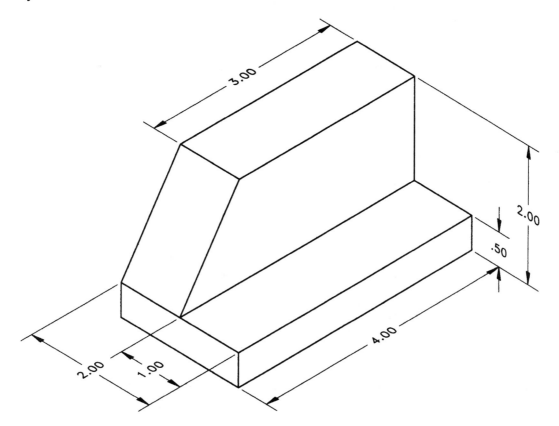

P2-9) Using a CAD package, draw the necessary views and completely dimension the objects given. Do not base your 2-D dimension placement on the 3-D dimensions shown. Use proper dimensioning techniques to dimension your object.

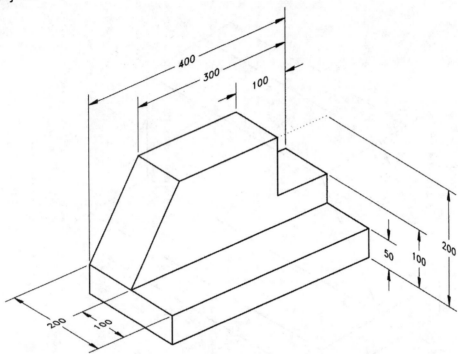

P2-10) Using a CAD package, draw the necessary views and completely dimension the objects given. Do not base your 2-D dimension placement on the 3-D dimensions shown. Use proper dimensioning techniques to dimension your object.

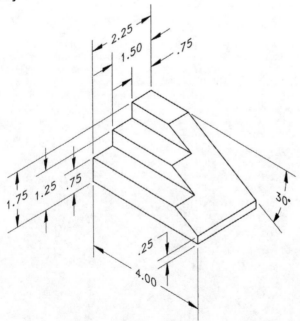

P2-11) Using a CAD package, draw the necessary views and completely dimension the objects given. Do not base your 2-D dimension placement on the 3-D dimensions shown. Use proper dimensioning techniques to dimension your object.

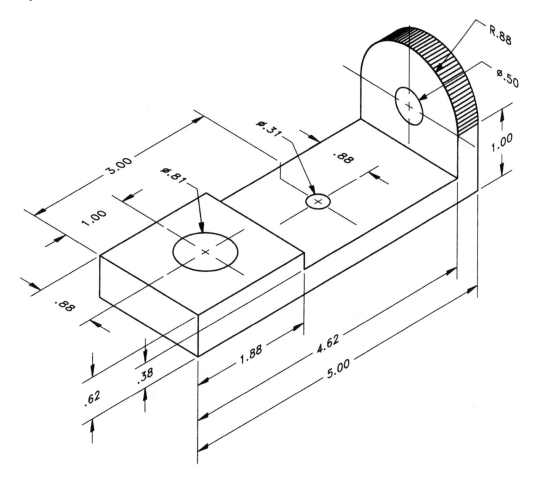

P2-12) Using a CAD package, draw the necessary views and completely dimension the objects given. Do not base your 2-D dimension placement on the 3-D dimensions shown. Use proper dimensioning techniques to dimension your object.

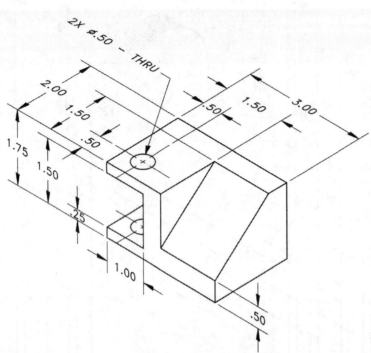

P2-13) Using a CAD package, draw the necessary views and completely dimension the objects given. Do not base your 2-D dimension placement on the 3-D dimensions shown. Use proper dimensioning techniques to dimension your object.

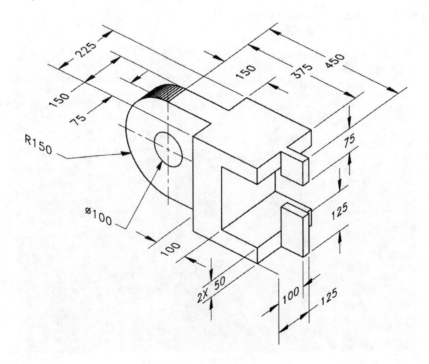

P2-14) Using a CAD package, draw the necessary views and completely dimension the objects given. Do not base your 2-D dimension placement on the 3-D dimensions shown. Use proper dimensioning techniques to dimension your object.

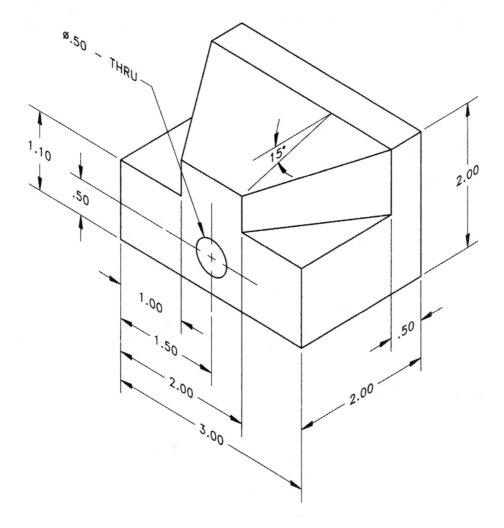

P2-15) Using a CAD package, draw the necessary views and completely dimension the objects given. Do not base your 2-D dimension placement on the 3-D dimensions shown. Use proper dimensioning techniques to dimension your object.

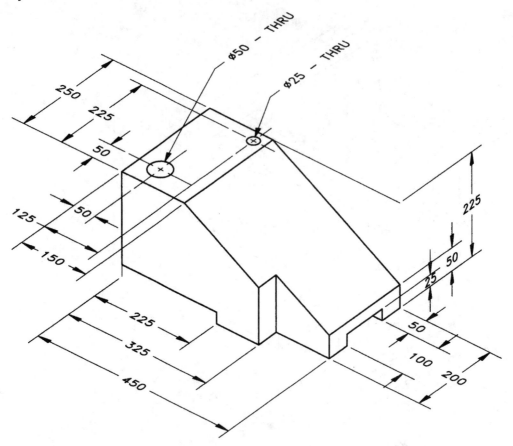

P2-16) Draw the necessary views and completely dimension the following objects. Do not base your 2-D dimension placement on the 3-D dimensions shown. Use proper dimensioning techniques to dimension your object.

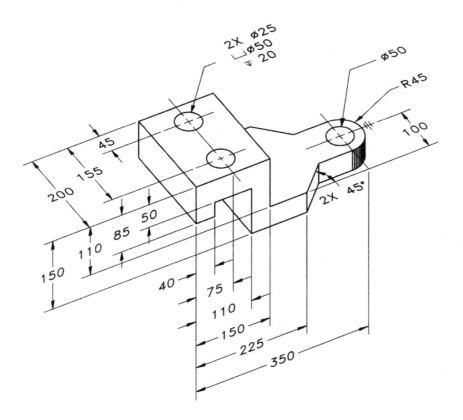

P2-17)   Draw the necessary views and completely dimension the following objects.   Do not base your 2-D dimension placement on the 3-D dimensions shown. Use proper dimensioning techniques to dimension your object.

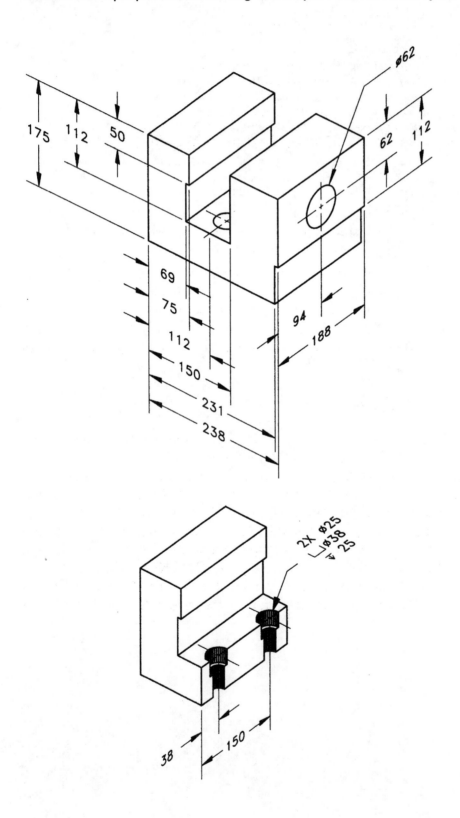

```
┌─────────────────────────┐
│      SECTIONING         │
└─────────────────────────┘
```

In Chapter 3 you will learn how to create various types of sectional views. Sectional views allow you to see inside an object. Using a sectional view within an orthographic projection can be very useful for parts that have complex interior geometry. By the end of this chapter, you will be able to create several different types of sectional views. You will also be able to choose which type of section is the most appropriate for a given part.

3.1) SECTIONAL VIEWS

A sectional view or section looks inside an object. Sections are used to clarify the interior construction of a part that can not be clearly described by hidden lines in exterior views. It is a cut away view of an object. Often, objects are more complex and interesting on the inside than on the outside. **By taking an imaginary cut through the object and removing a portion, the inside features may be seen more clearly.** For example, a geode is a rock that is very plain and featureless on the outside, but cut into it and you get an array of beautiful crystals.

3.1.1) Creating a Section View

To produce a section view, the part is cut using an imaginary cutting plane. The portion of the part that is between the observer and the cutting plane is mentally discarded exposing the interior construction as shown in Figure 3-1.

A sectional view should be projected perpendicular to the cutting plane and conform to the standard arrangement of views. If there are more than one section, they should be labeled with capital letters such as A, B or C. These letters are placed near the arrows of the cutting plane line. The sectional view is then labeled with the corresponding letter (e.g. SECTION A-A) as shown in Figure 3-2. Letters that should not be used to label sections are I, O, Q, S, X and Z. These letters may lead to misinterpretation. They are often used for other purposes.

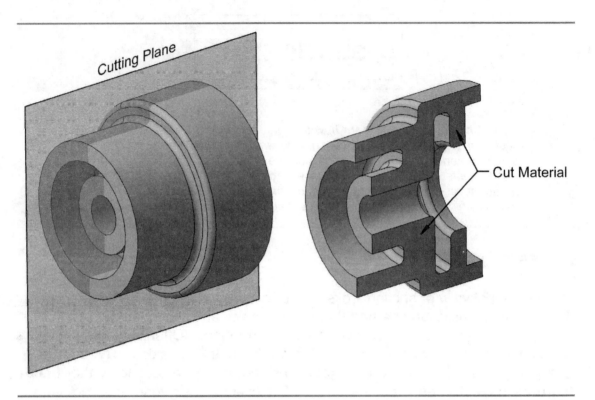

Figure 3-1: Creating a section view.

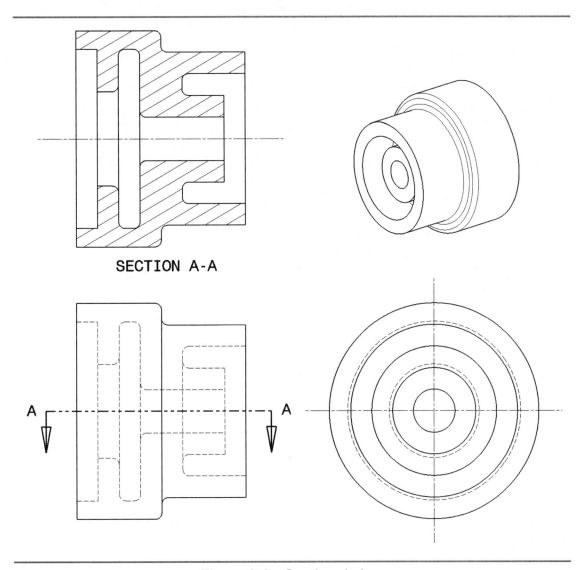

SECTION A-A

Figure 3-2: Sectional view.

3.1.2) <u>Lines used in Sectional Views</u>

• <u>Cutting Plane Line</u>

A cutting plane line is used to show where the object is being cut and represents the edge view of the cutting plane. Arrows are placed at the ends of the cutting plane line to indicate the direction of sight. The arrows point to the portion of the object that is kept. Cutting plane lines are thick (0.6 to 0.8 mm) and take precedence over centerlines. Figure 3-3 shows the two different types of cutting plane lines that are used on prints and Figure 3-2 illustrates its use.

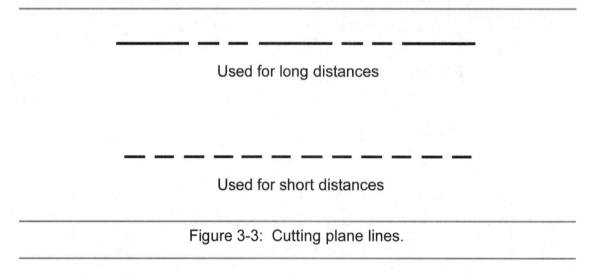

Used for long distances

Used for short distances

Figure 3-3: Cutting plane lines.

• <u>Section Lines</u>

Section lines are used to indicate where the cutting plane cuts the material (see Figure 3-2). Cut material is that which makes contact with the cutting plane. Section lines have the following properties:

√ Section lines are thin lines (0.3 mm).
√ Section line symbols (i.e. line type and spacing) are chosen according to the material from which the object is made. Figure 3-4 shows some of the more commonly used section line symbols.
√ Section lines are drawn at a 45° angle to the horizontal unless there is some advantage in using a different angle.

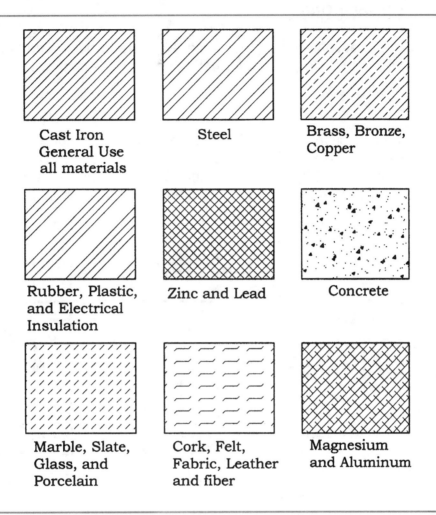

Figure 3-4: Section line symbols.

3.1.3) <u>Rules of Sectioning</u>

Rule 1. A section lined area is always completely bounded by a visible outline.

Rule 2. The section lines in all sectioned areas should be parallel. Section lines shown in opposite directions indicate a different part.

Rule 3. All the visible edges behind the cutting plane should be shown.

Rule 4. Hidden features should be omitted in all areas of a section view. Exceptions include threads and broken out sections.

3.2) <u>TYPES OF SECTIONS</u>

Many types of sectioning techniques are available to use. The type chosen depends on the situation and what information needs to be conveyed.

3.2.1) <u>Full Section</u>

To create a full section, the cutting plane passes fully through the object. The half of the object that is between the observer and the cutting plane is mentally removed exposing the cut surface and visible background lines of the remaining portion. Full sections are used in many cases to avoid having to dimension hidden lines as shown in Figure 3-5.

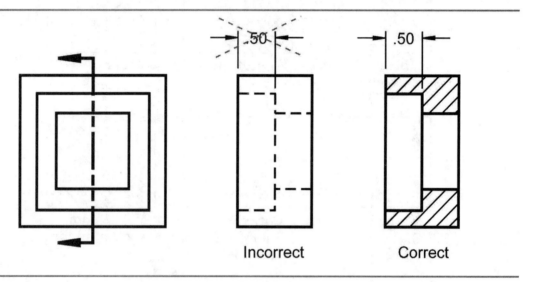

Figure 3-5: Full section.

Instructor Led Exercise 3-1: Full section

Given the top and right side views, sketch the front view as a full section. The material used is steel.

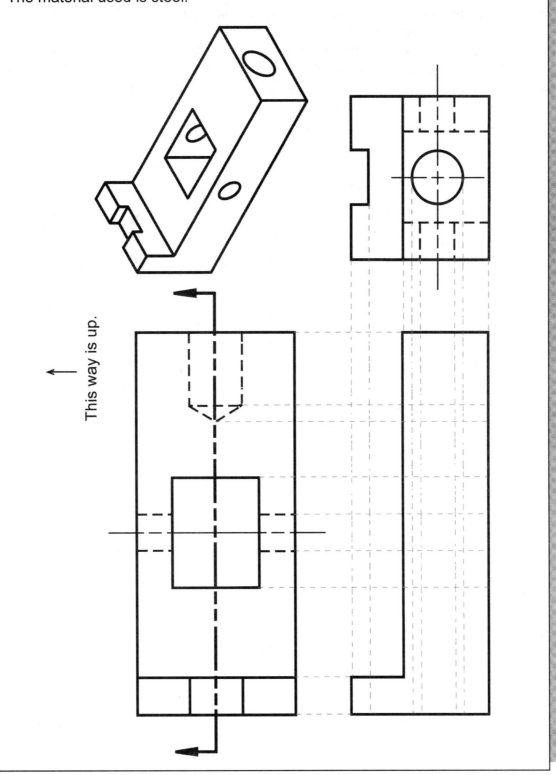

This way is up.

3.2.2) <u>Half Section</u>

A half section has the advantage of exposing the interior of one half of an object while retaining the exterior of the other half. Half sections are used mainly for symmetric or nearly symmetric objects or assembly drawings. The half section is obtained by passing two cutting planes through the object, at right angles to each other, such that the intersection of the two planes coincides with the axis of symmetry. Therefore, only a quarter of the object is mentally removed. On the sectional view, a centerline is used to separate the sectioned and unsectioned halves. Hidden lines should not be shown on either half. Figure 3-6 shows an example of a half section.

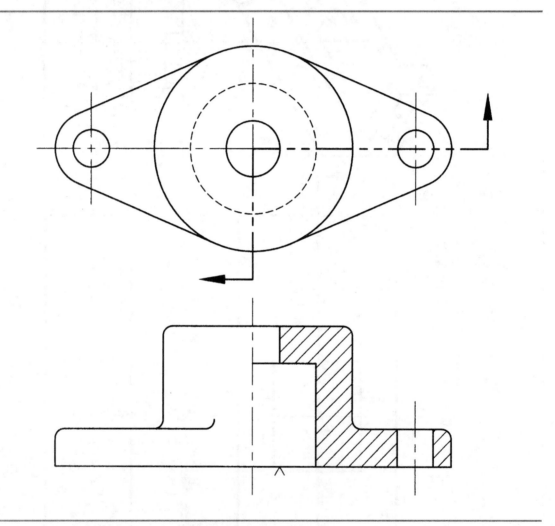

Figure 3-6: Half section.

<u>Instructor Led Exercise 3-2: Half section</u>

Given the front and right side views, sketch the top view as a full section and create a half sectioned front view. The material is brass.

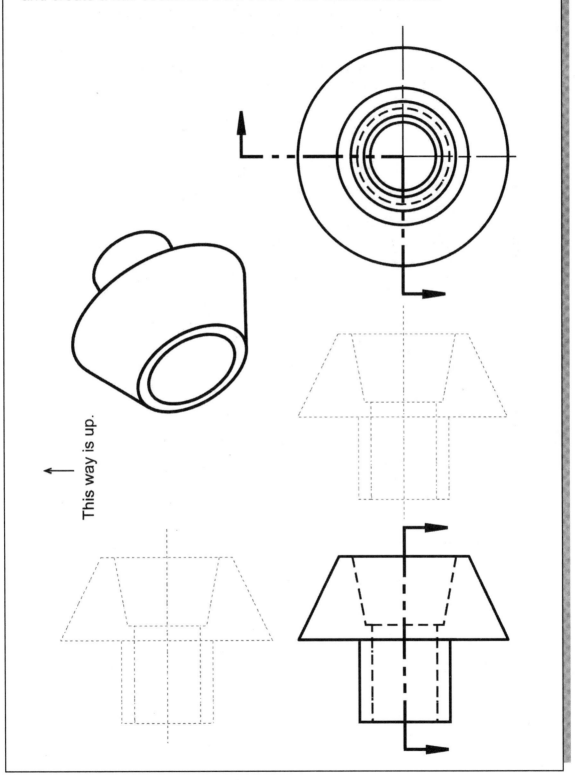

This way is up.

3.2.3) <u>Offset Section</u>

An offset section is produced by bending the cutting plane to show features that don't lie in the same plane. The section is drawn as if the offsets in the cutting plane were in one plane. Figure 3-7 shows an offset section.

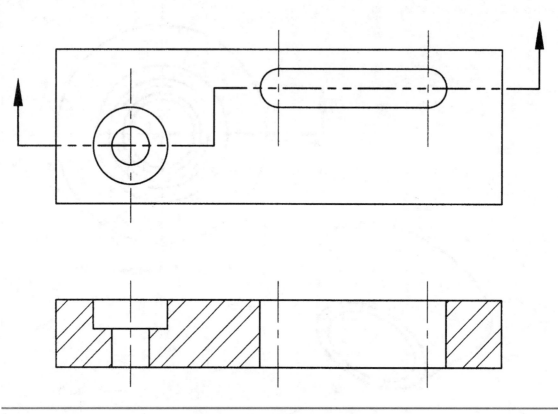

Figure 3-7: Offset section.

In Class Student Exercise 3-3: Offset section

Name: _____ Date: _____

Given the front and top views, sketch the three missing section views in their appropriate places. The material is cast iron.

SECTION C-C

SECTION B-B

SECTION A-A

This way is up.

A

B

C

NOTES:

3.2.4) <u>Aligned Section</u>

In order to include angled elements in a section, the cutting plane may be bent so that it passes through those features. The plane and features are then revolved, according to the conventions of revolution, into the original plane.

o <u>Conventions of Revolution:</u> Features are revolved into the projection plane, usually a vertical or horizontal plane, and then projected. The purpose of this is to show a true distance from a center or to show features that would otherwise not be seen. Figure 3-8 shows an aligned section employing the conventions of revolution.

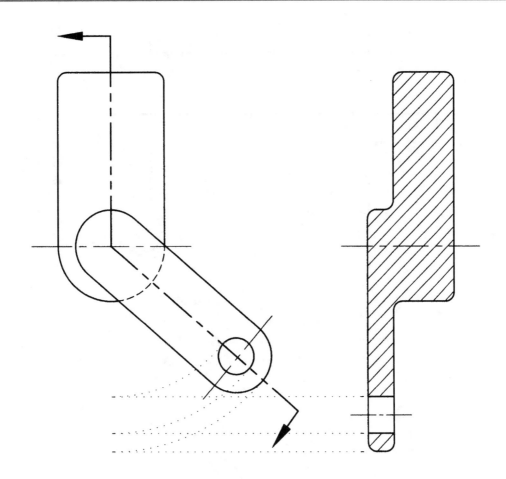

Figure 3-8: Aligned section.

Instructor Led Exercise 3-4: Aligned section

Given the front and unrevolved right side views, sketch the right side view as an aligned section using the conventions of revolution. The material is cast iron.

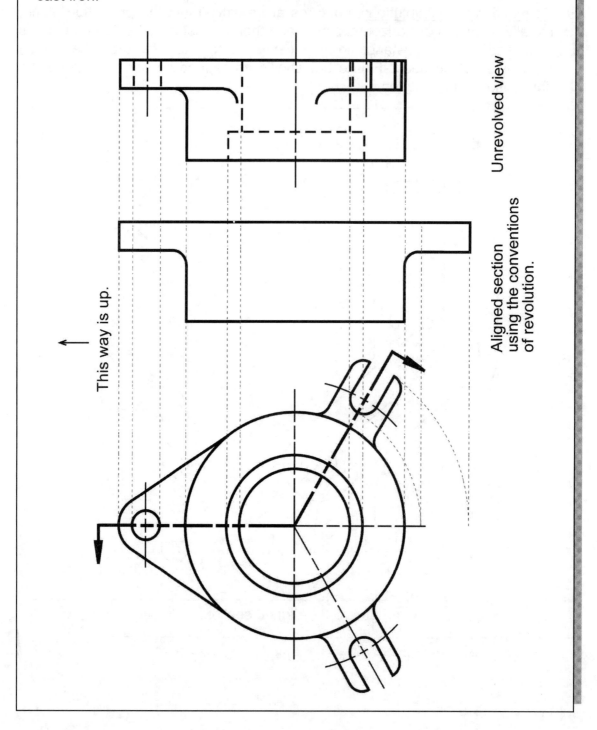

This way is up.

Unrevolved view

Aligned section using the conventions of revolution.

3.2.5) <u>Rib and Web Sections</u>

To avoid a false impression of thickness and solidity, ribs and webs and other similar features are not sectioned even though the cutting plane passes along the center plane of the rib or web. However, if the cutting plane passes crosswise through the rib or web, the member is shown in section as indicated in Figure 3-9.

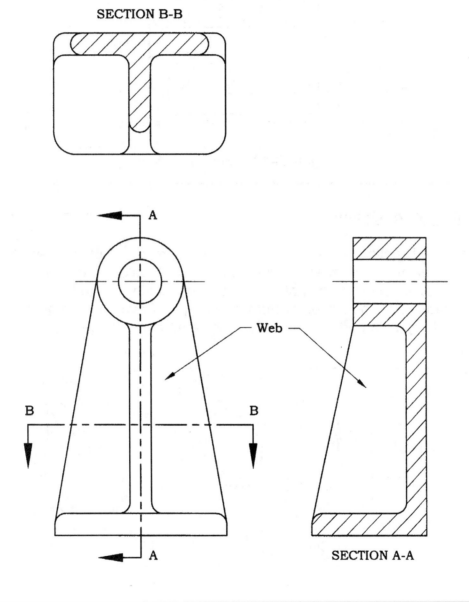

Figure 3-9: Rib and web sections.

3.2.6) Broken Section

Sometimes only a portion of the object needs to be sectioned to show a single feature of the part. In this case, the sectional area is bounded by a break line. Hidden lines are shown in the unsectioned area of a broken section. Figure 3-10 shows an example of a broken section.

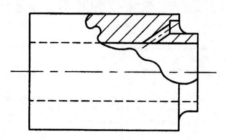

Figure 3-10: Broken section.

3.2.7) Removed Section

A removed section is one that is not in direct projection of the view containing the cutting plane. Removed sections should be labeled (e.g. SECTION A-A) according to the letters placed at the ends of the cutting plane line. They should be arranged in alphabetical order from left to right. Frequently, removed sections are drawn to an enlarged scale, which is indicated beneath the section title.

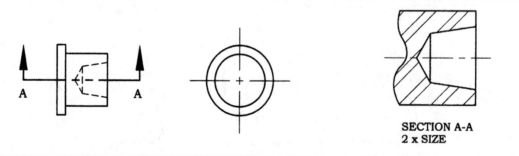

Figure 3-11: Removed section.

3.2.8) Revolved Section

The cross sectional shape of a bar, arm, spoke or other elongated objects may be shown in the longitudinal view by means of a revolved section. The visible lines adjacent to a revolved section may be broken out if desired. The super imposition of the revolved section requires the removal of all original lines covered by the section as shown in Figure 3-12. The true shape of a revolved section should be retained after the revolution regardless of the direction of the lines in the view.

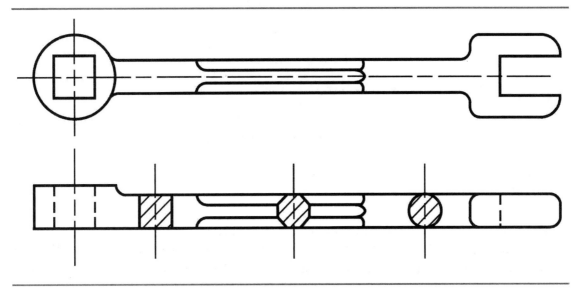

Figure 3-12: Revolved section.

3.2.9) Non-Sectioned Parts

It is common practice to show standard parts like nuts, bolts, rivets, shafts and screws 'in the round' or un-sectioned. This is done because they have no internal features. Other non-sectioned parts include bearings, gear teeth, dowels, and pins.

3.2.10) Thin Sections

For extremely thin parts of less than 4 mm thickness, such as sheet metal, washers, and gaskets, section lines are ineffective; therefore, the parts should be shown in solid black or without section lines.

NOTES:

SECTIONING REVIEW QUESTIONS

Name: _____   Date: _____

Answer the following questions.

Q3-1)   Explain the purpose of a section view.

Q3-2)   The arrows at the end of a cutting plane line point to the part of the object that is being (kept, thrown away).

Q3-3)   What does a cutting plane line indicate?

Q3-4)   What are section lines used to indicate?

Q3-5)   Section line symbols depend on (type of section, type of material).

Q3-6)   The sectioned and non-sectioned halves of a half section are separated by a (visible, center) line.

Q3-7)   A full section removes (¼, ½, ¾) of the object.

Q3-8)   A half section removes (¼, ½, ¾) of the object.

Q3-9)   Is it permissible to show hidden lines on some portion of a half section?

Q3-10)   Why do we use the conventions of revolution when creating an aligned section?

<u>NOTES:</u>

SECTIONING PROBLEMS

Name: _____ Date: _____

P3-1) Sketch the sectional view as indicated. The material of the part is steel.

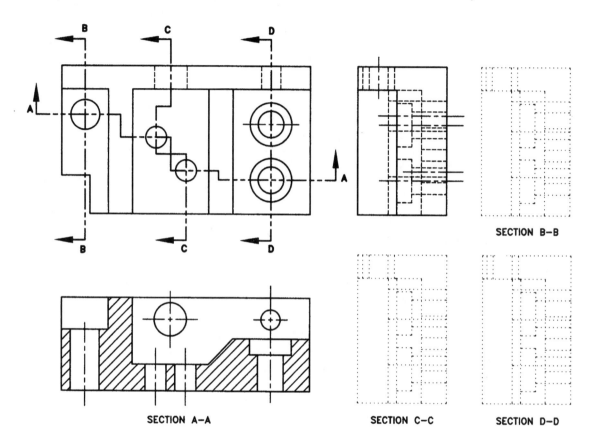

SECTION A-A

SECTION B-B

SECTION C-C

SECTION D-D

NOTES:

Name: _____   Date: _____

P3-2) Sketch the sectional view as indicated. The material of the part is cast iron.

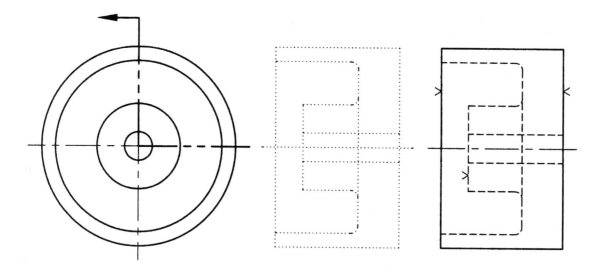

P3-3) Sketch the sectional view as indicated. The material of the part is plastic.

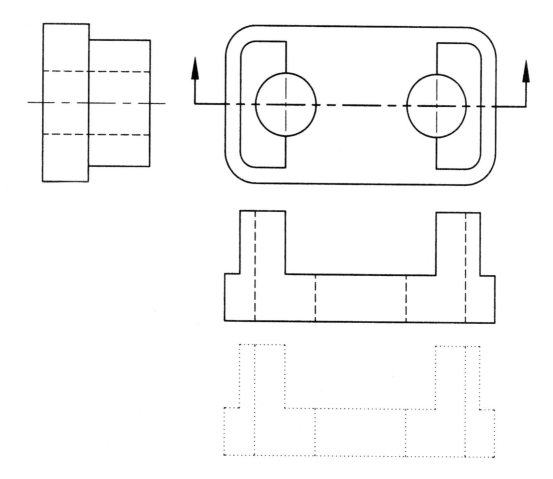

NOTES:

Name: _____ Date: _____

P3-4) Sketch the sectional view as indicated. The material of the part is cast iron.

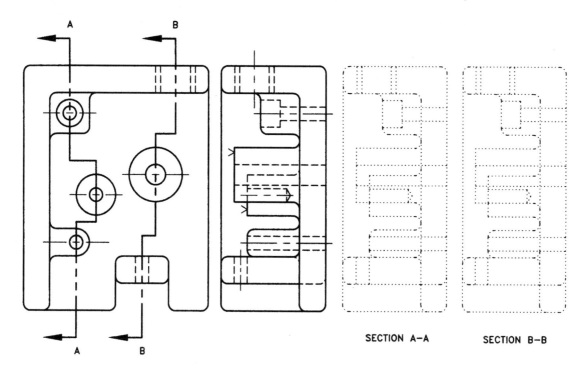

SECTION A-A   SECTION B-B

P3-5) Sketch the sectional view as indicated. The material of the part is cast iron.

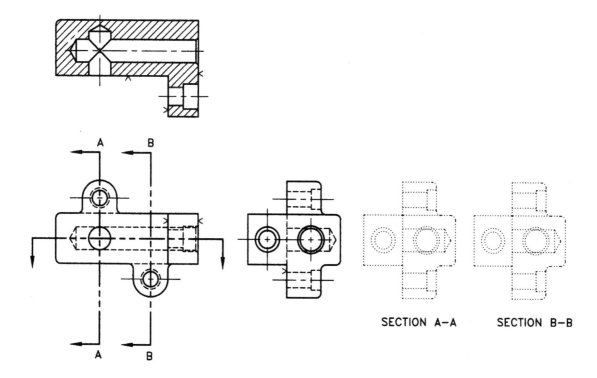

SECTION A-A   SECTION B-B

NOTES:

Name: _____ Date: _____

P3-6) Sketch the sectional view as indicated. The material of the part is brass.

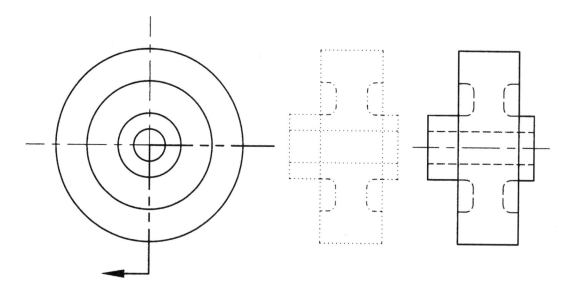

P3-7) Sketch the sectional view as indicated. The material of the part is steel.

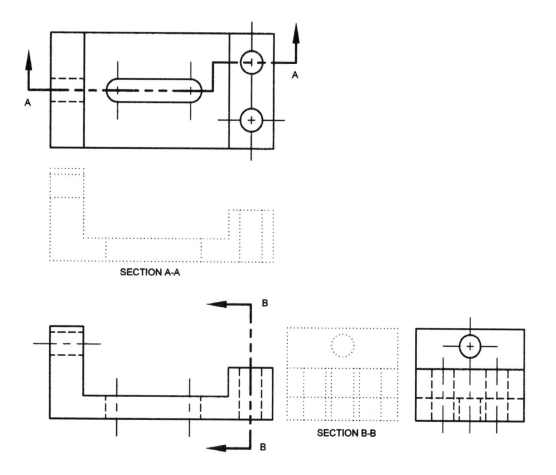

NOTES:

P3-8) Draw the following objects changing the appropriate view into a section. Do not dimension unless instructed to do so. The material of the part is steel.

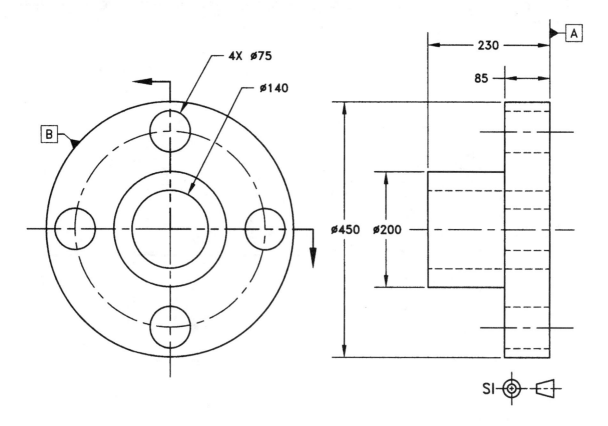

P3-9) Draw the following objects changing the appropriate view into a section. Do not dimension unless instructed to do so. The material of the part is cast iron. Note that without the section view it was necessary to dimension a hidden feature.

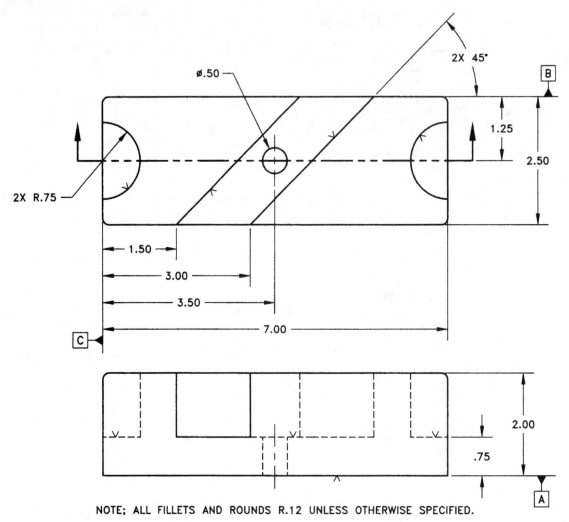

NOTE; ALL FILLETS AND ROUNDS R.12 UNLESS OTHERWISE SPECIFIED.

P3-10) Draw the following objects changing the appropriate view into a section. Do not dimension unless instructed to do so. The material of the part is aluminum.

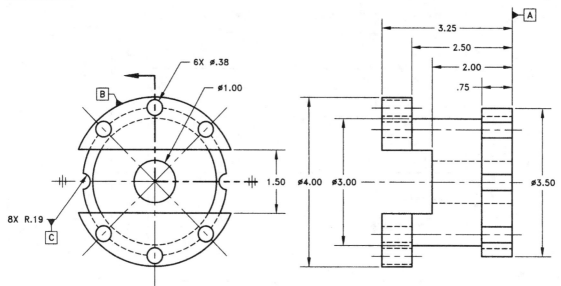

P3-11) Draw the following objects changing the appropriate view into a section. Do not dimension unless instructed to do so. The material of the part is steel.

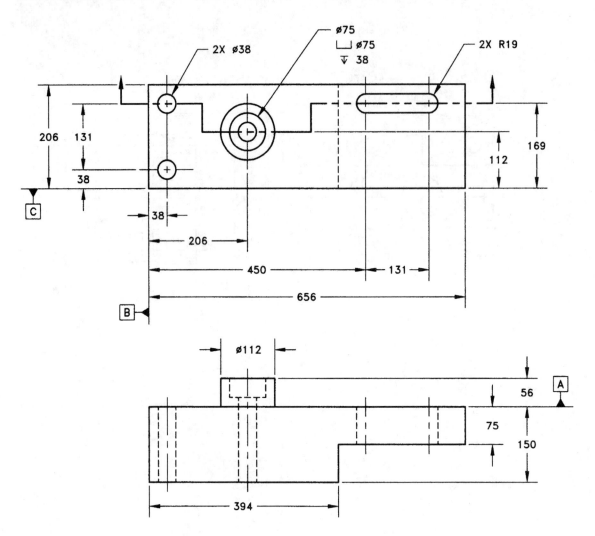

P3-12) Draw the following objects changing the appropriate view into a section. Do not dimension unless instructed to do so. The material of the part is aluminum.

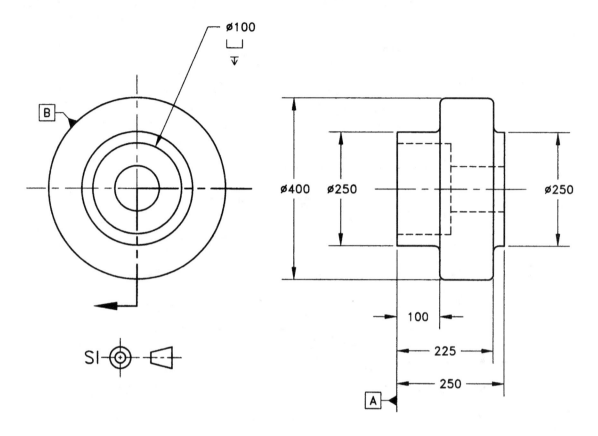

P3-13) Draw the following objects changing the appropriate view into a section. Do not dimension unless instructed to do so. The material of the part is plastic.

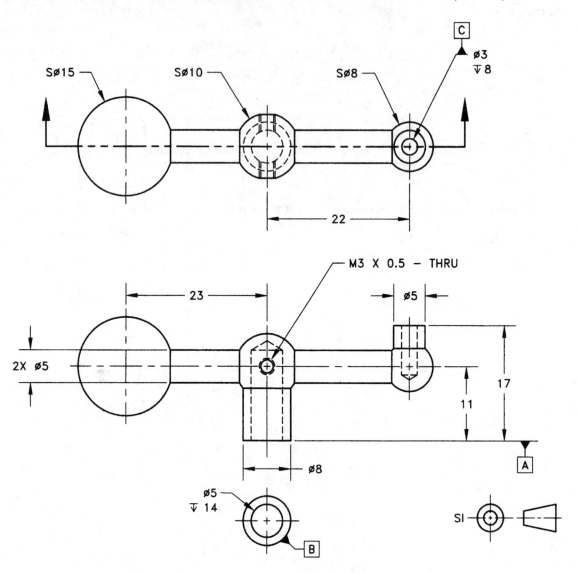

P3-14) Draw the following objects changing the appropriate view into a section. Do not dimension unless instructed to do so. The material of the part is steel.

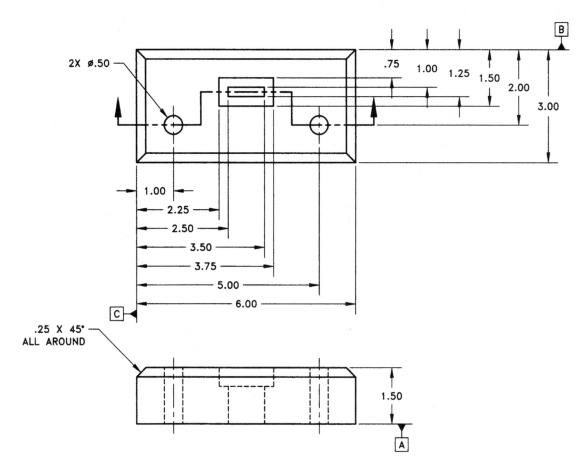

P3-15) Draw the following objects. Make use of the appropriate sectioning technique. Remember to indicate the cut plane. Do not dimension unless instructed to do so. The material of the part is steel.

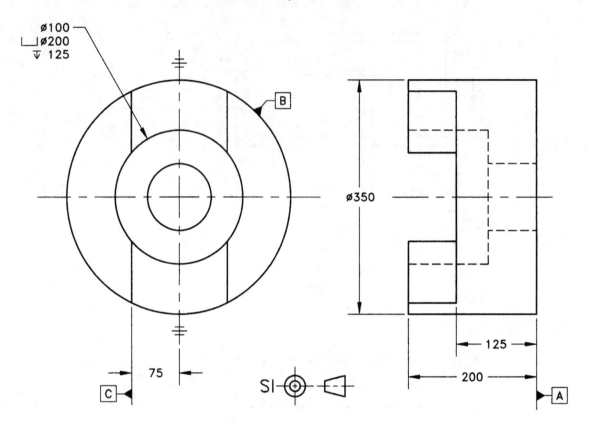

P3-16) Draw the following objects. Make use of the appropriate sectioning technique. Remember to indicate the cut plane. Do not dimension unless instructed to do so. The material of the part is aluminum.

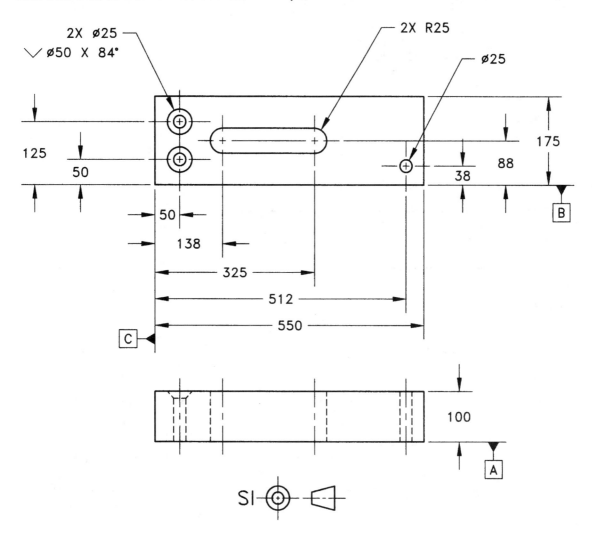

P3-17) Draw the following objects. Make use of the appropriate sectioning technique. Remember to indicate the cut plane. Do not dimension unless instructed to do so. The material of the part is steel.

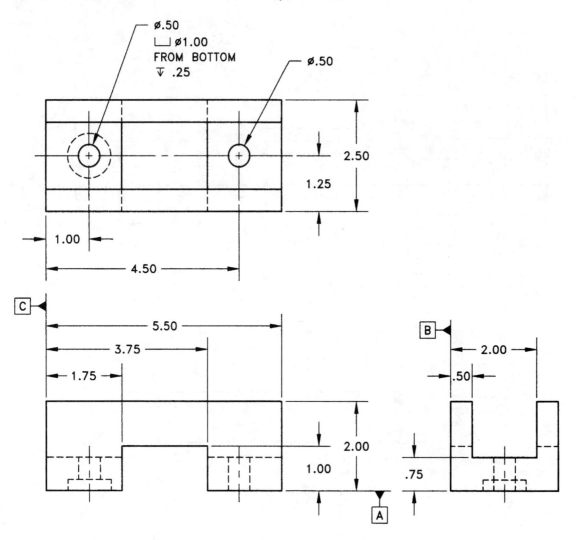

P3-18) Draw the following objects. Make use of the appropriate sectioning technique. Remember to indicate the cut plane. Do not dimension unless instructed to do so. The material of the part is steel.

2X ⌀1.00
⌴ ⌀1.75
▽ .38

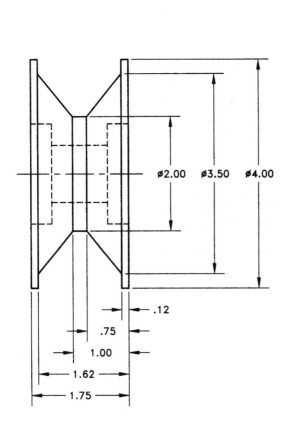

Ø2.00   Ø3.50   Ø4.00

.12
.75
1.00
1.62
1.75

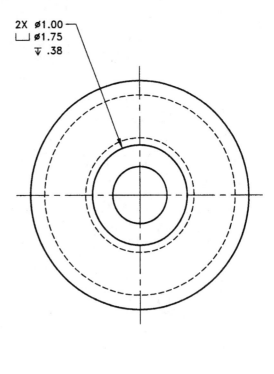

P3-19) Draw the following objects. Make use of the appropriate sectioning technique. Remember to indicate the cut plane. Do not dimension unless instructed to do so. The material of the part is brass.

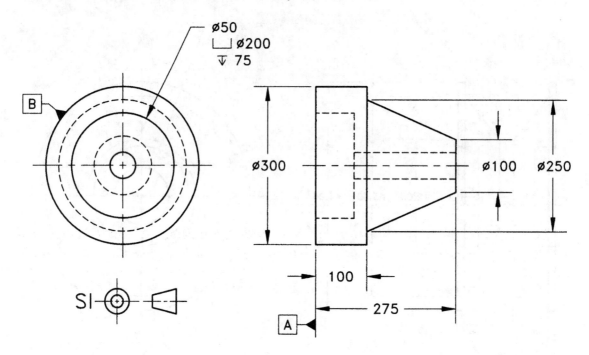

TOLERANCING

In Chapter 4 you will learn about tolerancing and how important this technique is to mass production. Tolerancing enables an engineer to design interchangeable or replacement parts. If a feature's size is toleranced, it is allowed to vary within a range of values or limits. It is no longer controlled by a single size. By the end of this chapter, you will be able to apply tolerances to a basic dimension and calculate a feature's limits.

4.1) TOLERANCING AND INTERCHANGEABILITY

Tolerancing is dimensioning for interchangeability. An interchangeable part is a part that possesses functional and physical characteristics equivalent in performance to another part for which it is intended to replace. When dimensioning an interchangeable part, the dimension is not a single value but a range of values that the part must fall within.

Interchangeability is achieved by imposing tolerances or limits on a dimension. **A tolerance is the total amount of dimensional variation permitted.** It is the amount that can be added to or subtracted from a basic size. Tolerancing enables similar parts to be near enough alike so that any one of them will fit properly into the assembly. For example, you would like to replace your mountain bike's seat post with a seat post that contains a shock absorber. You expect that all seat posts designed for mountain bikes will be interchangeable.

Tolerancing and interchangeability are an essential part of mass production. **Tolerances are necessary because it is impossible to manufacture parts without some variation.** A key component of mass production is the ability to buy replacement parts that are interchangeable or can substitute for the part being replaced.

Tolerancing gives us the means of specifying dimensions with whatever degree of accuracy we may require for our design to work properly. We would like to choose a tolerance that is not unnecessarily accurate or excessively inaccurate. Choosing the correct tolerance for a particular application depends on the design intent (end use) of the part, cost, how it is manufactured, and experience.

4.2) TOLERANCE TYPES

The tolerancing methods presented in this chapter include *limit* dimensions, *plus or minus* tolerances, and *page* or *block* tolerances.

- Limit Dimensions: Limits are the maximum and minimum size that a part can obtain. For example, the diameter of a shaft may vary between .999 inch and 1.001 inches. On a drawing, you would see this dimension specified as one of the following:

$$\varnothing 1.001_{.999} \quad \text{or} \quad \varnothing 1.001 - .999$$

Limit dimensions provide the blueprint reader with the limits of allowable variation without any calculation. It will eliminate potential calculation mistakes. Limits for external dimensions are generally given with the larger dimension first or on top and the smaller dimension last or on the bottom. Limits for internal dimensions are generally given with the smaller dimension first and the larger dimension last. This convention is mainly used to avoid machining mistakes.

- Plus or Minus Tolerances: Plus or minus tolerances give a basic size and the variation that can occur around that basic size. On a drawing, a plus or minus tolerance dimension would look like the following:

$$10.0 \, {}^{+0.1}_{-0.2}$$

When the positive and negative variations are the same, it is referred to as an equal bilateral tolerance. When they are not the same, the specification is called an unequal bilateral and when one of the variances is zero, the tolerance is unilateral. The type of tolerance chosen depends on the direction in which variation is most detrimental. Plus or minus tolerances are convenient because a design may be initially drawn and dimensioned using basic sizes. As the design progresses, tolerances may be added.

- Page or Block Tolerances: Page tolerances, also called block tolerances, get their name from their location on the drawing. Page tolerances are generally placed in the lower right hand corner of the page in or near the title block. The page tolerance is actually a general note that applies to all dimensions not covered by some other tolerancing type. The format of a page tolerance note may look like the following.

UNLESS OTHERWISE SPECIFIED ALL:
.XX = ± .010 inch
.XXX = ± .005 inch
.XXXX = ± .002 inch

Page tolerances are used for two reasons. First, they act as a default tolerance for any dimensions that may have been overlooked when tolerances were assigned. Second, they are often used to quickly tolerance non-critical dimensions.

4.3) **GENERAL DEFINITIONS**

In the intervening sections, a simple shaft and hole assembly will be used to illustrate different concepts and definitions. Figure 4-1 shows a shaft that is designed to fit into a hole. Both the shaft and the hole are allowed to vary between a maximum and minimum diameter.

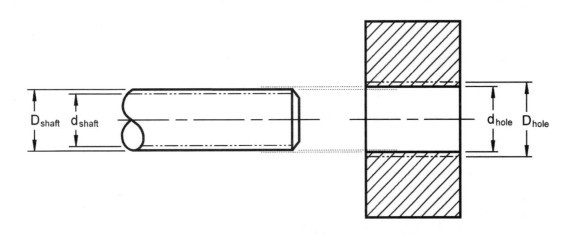

Figure 4-1: Shaft and hole assembly.

The terms *limits*, *tolerance* and *allowance* are often used interchangeably, but each of these three terms refers to something different.

- Limits: The limits are a feature's allowable maximum and minimum size.

- Tolerance: The tolerance is the difference between the maximum and minimum size of a feature.

- Allowance: The allowance is the difference between the smallest size of the hole and the largest size of the shaft. It may be positive or negative. The term allowance and minimum clearance are often used interchangeably.

Instructor Led Exercise 4-1: General definitions

Referring to Figure 4-1, answer the following questions.

- What are the limits of the shaft and the hole?

    Shaft:

    Hole:

- What is the tolerance for the shaft and the hole?

    Shaft:

    Hole:

- What is the minimum clearance (allowance)?

- What is the maximum clearance?

4.4) TOLERANCING STANDARDS

Standards are needed to establish dimensional limits for parts that are to be interchangeable. Standards make it possible for parts to be manufactured at different times and in different places with the assurance that they will meet assembly requirements. The two most common standards agencies are the American National Standards Institute (ANSI) and the International Standards Organization (ISO). The ANSI standards are now being compiled and distributed by the American Society of Mechanical Engineers (ASME). The information contained in this chapter is based on the following standards: ASME Y14.5 - 1994, ANSI B4.1 – 1967 (R1994), and ANSI B4.2 – 1978 (R1984).

4.5) INCH TOLERANCES

The simple shaft and hole assembly, shown in Figure 4-2, will be used to illustrate different concepts and definitions relating to tolerancing in inches. The dimensions of both are shown in Figure 4-2. Both diameters are allowed to vary between a maximum and minimum value.

- Basic Size: The basic size is the size from which the limits are calculated. It is common for both the hole and the shaft and is usually the closest fraction.

- Limits: The limits are the maximum and minimum size that the part is allowed to be.

- Tolerance: The tolerance is the total amount a specific dimension is permitted to vary.

- Maximum Material Condition (MMC): The MMC is the size of the part when it consists of the most material.

- Least Material Condition (LMC): The LMC is the size of the part when it consists of the least material.

- Maximum Clearance: The maximum clearance is the maximum amount of space that can exist between the hole and the shaft. The maximum clearance is calculated by using the following equation:

$$\text{Max. Clearance} = \text{LMC}_{hole} - \text{LMC}_{shaft}$$

- Minimum Clearance (Allowance): The minimum clearance is the minimum amount of space that can exist between the hole and the shaft. The minimum clearance is calculated by using the following equation:

$$\text{Min. Clearance} = \text{MMC}_{hole} - \text{MMC}_{shaft}$$

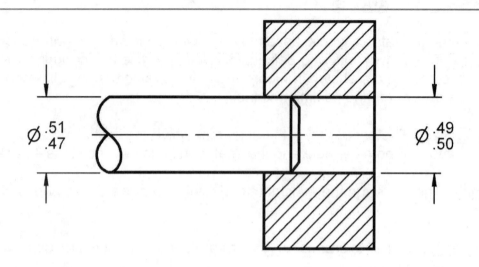

Figure 4-2: Toleranced shaft and hole pair (English).

Instructor Led Exercise 4-2: Inch tolerance definitions

Referring to Figure 4-2, fill in the following table.

	Shaft	Hole
Limits		
Basic Size		
Tolerance		
MMC		
LMC		
Max. Clearance		
Min. Clearance (Allowance)		

4.5.1) Types of Fits

Within the set of inch tolerances, there are four major types of fits. Within each fit there are several degrees or classes. The type of fit and class that you choose to implement depends on the function of your design. The major categories of fits are: *clearance*, *interference*, *transition*, and *line*. Table 4-1 lists and defines each of the four types of fits.

Type of Fit	Definition	When it exists
Clearance Fit	The internal member (shaft) fits into the external member (hole) and always leaves a space or clearance between the parts.	Min. Clear > 0
Interference Fit	The internal member is larger than the external member such that there is always an actual interference of metal.	Max. Clear ≤ 0
Transition Fit	The fit might result in either a clearance or interference fit condition.	Min. Clear < 0 Max. Clear > 0
Line Fit	The limits of size are specified such that a clearance or surface contact may result.	Min. Clear = 0 Max. Clear >0

Table 4-1: Types of fits.

Instructor Led Exercise 4-3: Types of fits

From everyday life, list some examples of clearance and interference fits.

Fit	Examples
Clearance	
Interference	

Instructor Led Exercise 4-4: Determining fit type

Determine the basic size and type of fit given the limits for the shaft and hole.

Shaft Limits	Hole Limits	Basic Size	Type of Fit
1.500 – 1.498	1.503 – 1.505		
.755 - .751	.747 - .750		
.378 - .373	.371 - .375		
.250 - .247	.250 - .255		

4.5.2) ANSI Standard Limits and Fits (English)

The following fit types and classes are in accordance with the ANSI B4.1-1967 (R1994) standard.

- Running or Sliding Clearance Fits (RC)

Running and sliding clearance fits are intended to provide running performance with suitable lubrication. Table 4-2 lists the different classes of running and sliding clearance fits and their design uses.

- Locational Fits (LC, LT, LN)

Locational fits are intended to determine only the location of the mating parts. They are divided into three groups: clearance fits (LC), transition fits (LT), and interference fits (LN). Table 4-3 lists the different classes of locational fits and their design uses.

- FN: Force Fits:

Force fits provide a constant bore pressure throughout the range of sizes. The classes are categorized from FN1 to FN5. Table 4-4 lists the different classes of force fits and their design uses.

Class of Fit	Description	Design use
RC9 -RC8	Loose running fit	Used with material such as cold rolled shafting and tubing made to commercial tolerances.
RC7	Free running fit	Used where accuracy is not essential, or where large temperature variations occur.
RC6 -RC5	Medium running fit	Used on accurate machinery with higher surface speeds where accurate location and minimum play is desired.
RC4	Close running fit	Used on accurate machinery with moderate surface speeds where accurate location and minimum play is desired.
RC3	Precision running fit	This is the closest fit, which can be expected to run freely. Intended for slow speeds. Not suitable for appreciable temperature changes.
RC2	Sliding fit	Used for accurate location. Parts will move and turn easily but are not intended to run freely. Parts may seize with small temperature changes.
RC1	Close Sliding fit	Used for accurate location of parts that must be assembled without perceptible play.

Table 4-2: Running and sliding clearance fit classes.

Class of Fit	Description	Design use
LC	Locational clearance fit	Intended for parts that are normally stationary, but which can be freely assembled or disassembled. They run from snug fits (parts requiring accuracy of location), through the medium clearance fits (parts where freedom of assembly is important). The classes are categorized from LC1 being the tightest fit to LC11 being the loosest.
LT	Locational transition fit	Used where accuracy of location is important, but a small amount of clearance or interference is permissible. The classes are categorized from LT1 to LT6.
LN	Locational interference fit	Used where accuracy of location is of prime importance, and for parts requiring rigidity and alignment with no special requirements for bore pressure. The classes are categorized from LN1 to LN3.

Table 4-3: Locational fit classes.

Class of Fit	Description	Design use
FN1	Light drive fit	This fit produces a light assembly pressure and, a more or less, permanent assembly.
FN2	Medium drive fit	Suitable for ordinary steel parts or for shrink fits on light sections. About the tightest fit that can be used with high-grade cast iron.
FN3	Heavy drive fit	Suitable for heavier steel parts or for shrink fit in medium sections.
FN4 - FN5	Force fit	Suitable for parts that can be heavily stressed, or for shrink fits where the heavy pressing forces required are impractical.

Table 4-4: Force fit classes.

Instructor Led Exercise 4-5: Limits and fits

Given a basic size of .50 inches and a fit of RC8, calculate the limits for both the hole and the shaft. Use the ANSI limits and fit tables given in the Appendix D.

Shaft:

Hole:

In Class Student Exercise 4-6: Milling Jack assembly tolerances

Name: _____ Date: _____

Consider the *Milling Jack* assembly shown. Notice that there are many parts that fit into or around other parts. Each of these parts are toleranced to ensure proper fit and function.

Milling Jack assembled and exploded views.

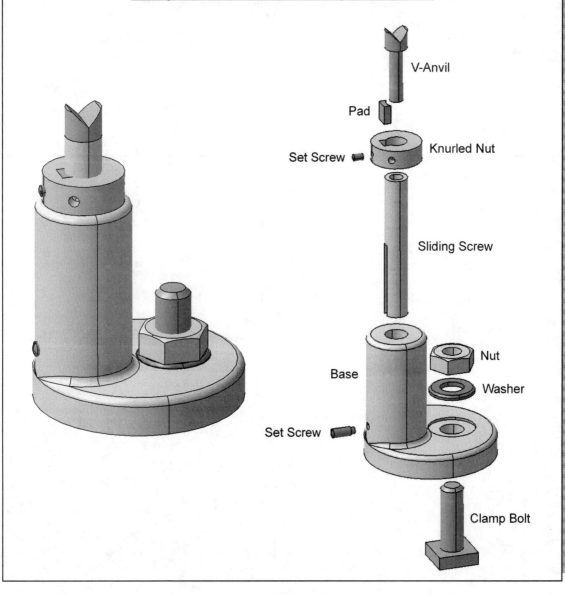

Milling Jack assembly drawing

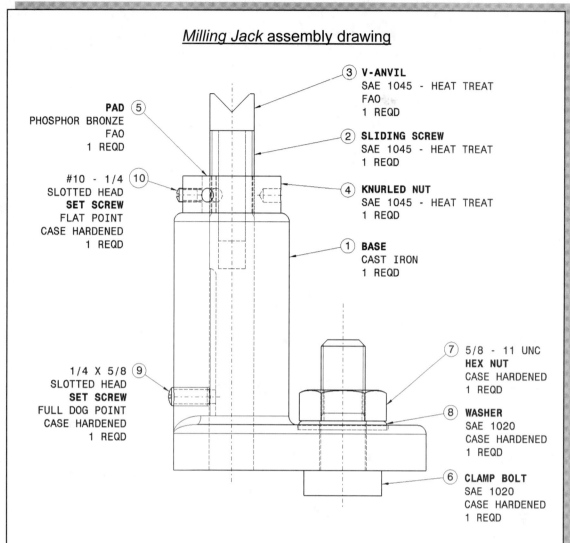

(3) **V-ANVIL**
SAE 1045 - HEAT TREAT
FAO
1 REQD

PAD (5)
PHOSPHOR BRONZE
FAO
1 REQD

(2) **SLIDING SCREW**
SAE 1045 - HEAT TREAT
1 REQD

#10 - 1/4 (10)
SLOTTED HEAD
SET SCREW
FLAT POINT
CASE HARDENED
1 REQD

(4) **KNURLED NUT**
SAE 1045 - HEAT TREAT
1 REQD

(1) **BASE**
CAST IRON
1 REQD

(7) 5/8 - 11 UNC
HEX NUT
CASE HARDENED
1 REQD

1/4 X 5/8 (9)
SLOTTED HEAD
SET SCREW
FULL DOG POINT
CASE HARDENED
1 REQD

(8) **WASHER**
SAE 1020
CASE HARDENED
1 REQD

(6) **CLAMP BOLT**
SAE 1020
CASE HARDENED
1 REQD

A dimensioned isometric drawing for each relevant part is given in the following pages.

- The *V-Anvil* fits into the *Sliding Screw* (see assembly drawing) with a RC4 fit. The basic size is .375 (3/8). Determine the limits for both parts.

- The *Sliding Screw* fits into the *Base* (see assembly drawing) with a RC5 fit. The basic size is .625 (5/8). Determine the limits for both parts.

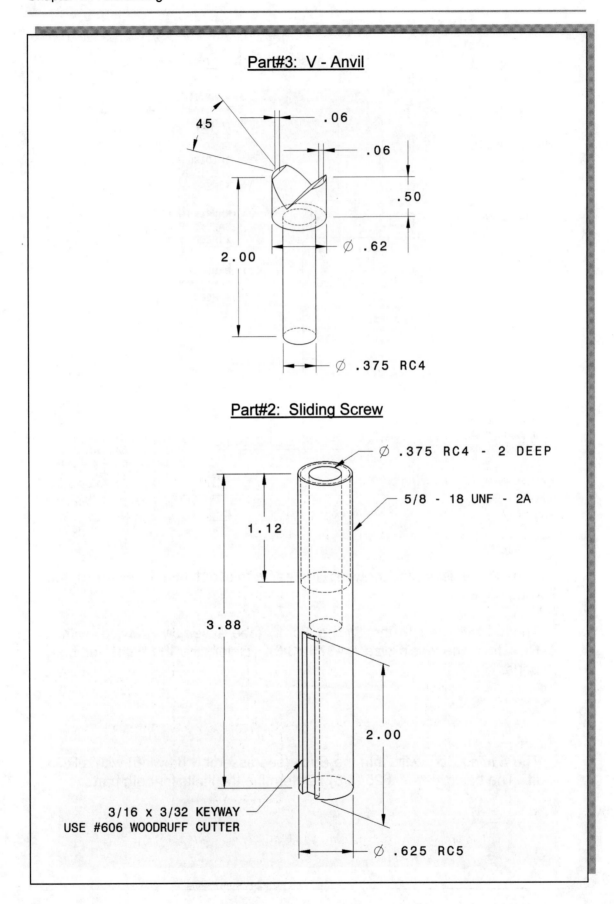

Part#3: V - Anvil

45
.06
.06
.50
2.00
Ø .62
Ø .375 RC4

Part#2: Sliding Screw

Ø .375 RC4 - 2 DEEP
5/8 - 18 UNF - 2A
1.12
3.88
2.00
3/16 x 3/32 KEYWAY
USE #606 WOODRUFF CUTTER
Ø .625 RC5

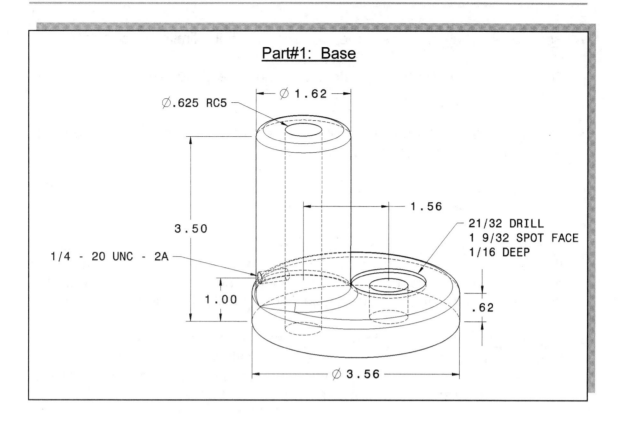

Part#1: Base

Ø.625 RC5

Ø 1.62

1.56

3.50

21/32 DRILL
1 9/32 SPOT FACE
1/16 DEEP

1/4 - 20 UNC - 2A

1.00

.62

Ø 3.56

4.6) <u>METRIC TOLERANCES</u>

The simple shaft and hole assembly, shown in Figure 4-3, will be used to illustrate different concepts and definitions relating to tolerancing in millimeters. The dimensions of both are shown in Figure 4-3. Both diameters are allowed to vary between a maximum and minimum value.

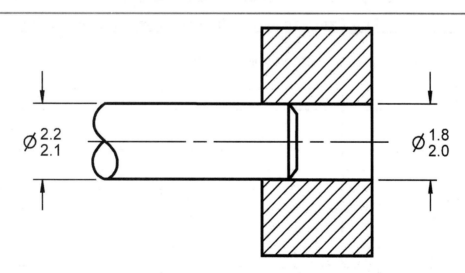

$\emptyset^{2.2}_{2.1}$ $\emptyset^{1.8}_{2.0}$

Figure 4-3: Toleranced shaft and hole pair (metric).

- <u>Basic Size:</u> The basic size is the size from which the limits are calculated.

- <u>Tolerance:</u> The tolerance is the total amount a dimension is permitted to vary.

- <u>Upper deviation:</u> The upper deviation is the difference between the basic size and the permitted maximum size of the part.

- <u>Lower deviation:</u> The lower deviation is the difference between the basic size and the minimum permitted size of the part.

- <u>Fundamental deviation:</u> The fundamental deviation is the closest deviation to the basic size. To determine the fundamental deviation, compare the upper deviation and the lower deviation. The fundamental deviation is the smaller of the two. A letter in the fit specification represents the fundamental deviation.

- <u>International tolerance grade number (IT#):</u> The IT#'s are a set of tolerances that vary according to the basic size and provide the same relative level of accuracy within a given grade. The number in the fit specification represents the IT#. A smaller number provides a smaller tolerance.

- <u>Tolerance zone:</u> The fundamental deviation in combination with the IT# defines the tolerance zone. The IT# establishes the magnitude of the tolerance zone or the amount that the dimension can vary. The fundamental deviation establishes the position of the tolerance zone with respect to the basic size.

Instructor Led Exercise 4-7: Millimeter tolerance definitions

Referring to Figure 4-3, fill in the following table.

	Shaft	Hole
Limits		
Basic Size		
Tolerance		
Upper deviation		
Lower deviation		
Fundamental deviation		
Type of fit		

4.6.1) <u>ANSI Standard Limits and Fits</u> (Metric)

The following fit types are in accordance with the ANSI B4.2-1978 (R1994) standard. Available metric fits and their descriptions are summarized in Table 4-5.

FIT SYMBOL			
Hole Basis	**Shaft Basis**	**Fit**	**Description**
H11/c11	C11/h11	Loose running fit	For wide commercial tolerances or allowances.
H9/d9	D9/h9	Free running fit	Good for large temperature variations, high running speeds, or heavy journal pressures.
H8/f7	F8/h7	Close running fit	For accurate location at moderate speeds and journal pressures.
H7/g6	G7/h6	Sliding fit	Not intended to run freely, but to move and turn freely and locate accurately.
H7/k6	K7/h6	Locational clearance fit	For locating stationary parts but can be freely assembled and disassembled.
H7/n6	N7/h6	Locational transition fit	For accurate location.
H7/p6	P7/h6	Locational interference fit	For parts requiring rigidity where accuracy of location is important, but without special bore pressure requirements.
H7/s6	S7/h6	Medium drive fit	For ordinary steel parts or shrink fits on light sections, the tightest fit usable with cast iron.
H7/u6	U7/h6	Force fit	Suitable for parts that can be highly stressed.

Table 4-5: Metric standard fits.

4.6.2) <u>Tolerance Designation</u>

Metric fits are specified using the fundamental deviation (letter) and the IT#. When specifying the fit for the hole, an upper case letter is used. A lower case letter is used when specifying the fit for the shaft. As stated before, the IT# establishes the magnitude of the tolerance zone or the amount that the dimension can vary. The fundamental deviation establishes the position of the tolerance zone with respect to the basic size.

Instructor Led Exercise 4-8: Metric fit designation

Fill in the appropriate name for the fit component.

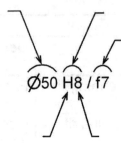

4.6.3) Basic Hole and Basic Shaft Systems

Notice that Table 4-5 gives two different tolerance designations for each type of fit. Metric limits and fits are divided into two different systems: the basic hole system and the basic shaft system. Each system has its own designation.

- Basic hole system: The basic hole system is used when you want the basic size to be attached to the hole dimension. For example, if a standard drill, reamer, broach, or another standard tool is used to produce a hole, you would want to use the hole system. In this system, the minimum hole diameter is taken as the basic size. This system is used to increase tool life.

- Basic shaft system: The basic shaft system is used when you want the basic size to be attached to the shaft dimension. For example, you would use the shaft system if you need to tolerance a hole based on the size of a purchased standard drill rod. In this system, the maximum shaft diameter is taken as the basic size.

Instructor Led Exercise 4-9: Systems

Identify the type of fit and the system used to determine the limits of the following shaft and hole pairs.

Shaft	Hole	Type of Fit	System
9.987 – 9.972	10.000 – 10.022		
60.021 - 60.002	60.000 - 60.030		
40.000 – 39.984	39.924 – 39.949		

Instructor Led Exercise 4-10: Metric limits and fits.

Find the limits, tolerance, type of fit, and type of system for a ⌀30 H11/c11 fit. Use the tolerance tables given in Appendix D.

	Shaft	Hole
Limits		
Tolerance		
System		
Fit		

Find the limits, tolerance, type of fit, and type of system for a ⌀30 P7/h6 fit.

	Shaft	Hole
Limits		
Tolerance		
System		
Fit		

4.7) SELECTING TOLERANCES

Tolerances will govern the method of manufacturing. **When tolerances are reduced, the cost of manufacturing rises very rapidly.** Therefore, specify as generous a tolerance as possible without interfering with the function of the part.

Choosing the most appropriate tolerance depends on many factors, such as length of engagement, bearing load, speed, lubrication, temperature, humidity, and material. Experience also plays a significant role.

Table 4-6 may be used as a general guide for determining the machining processes that will under normal conditions, produce work within the tolerance grades indicated. As the tolerance grade number decreases the tolerance becomes smaller. Tolerance grades versus actual tolerances may be found in any Machinery's Handbook.

Machining Operation	IT Grades							
	4	5	6	7	8	9	10	11
Lapping & Honing	▓	▓						
Cylindrical Grinding		▓	▓	▓				
Surface Grinding		▓	▓	▓				
Diamond Turning		▓	▓	▓				
Diamond Boring		▓	▓	▓				
Broaching		▓	▓	▓				
Reaming			▓	▓	▓	▓		
Turning				▓	▓	▓	▓	
Boring				▓	▓	▓	▓	
Milling						▓	▓	▓
Planing & Shaping						▓	▓	▓
Drilling						▓	▓	▓
Punching							▓	▓
Die Casting							▓	▓

Table 4-6: Relation of machining processes to international tolerance grades.

4.8) TOLERANCE ACCUMULATION

The tolerance between two features of a part depends on the number of controlling dimensions. The distance could be controlled by a single dimension or multiple dimensions. The maximum variation between two features is equal to the sum of the tolerances placed on the intermediate dimensions. As the number of intermediate dimensions increases, the tolerance accumulation increases. Remember, even if the dimension does not have a stated tolerance, it has an implied tolerance.

Instructor Led Exercise 4-11: Tolerance accumulation

What is the tolerance accumulation for the distance between surface A and B for the three different dimensioning methods?

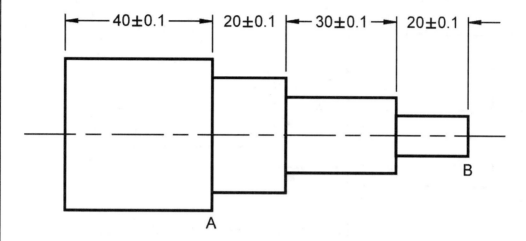

Tolerance accumulation between surface A and B =

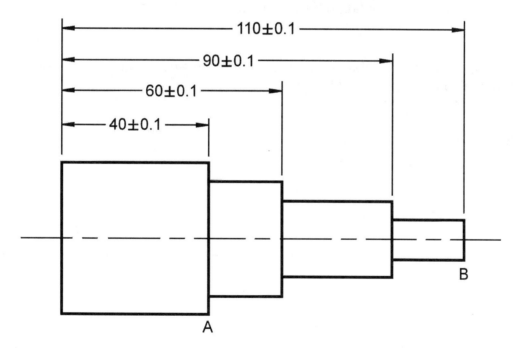

Tolerance accumulation between surface A and B =

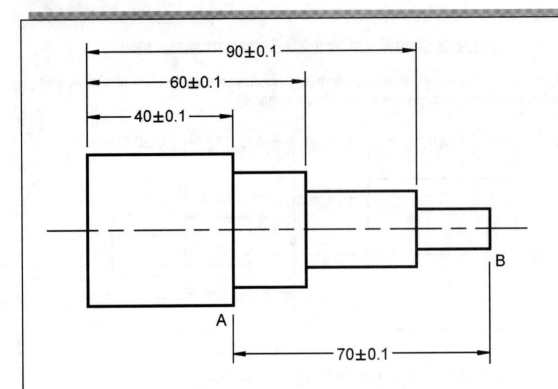

Tolerance accumulation between surface A and B =

If the accuracy of the distance between surface A and B is important, which dimensioning method should be used?

Instructor Led Exercise 4-12: Over dimensioning

Assuming that the diameter dimensions are correct, explain why this object is dimensioned incorrectly.

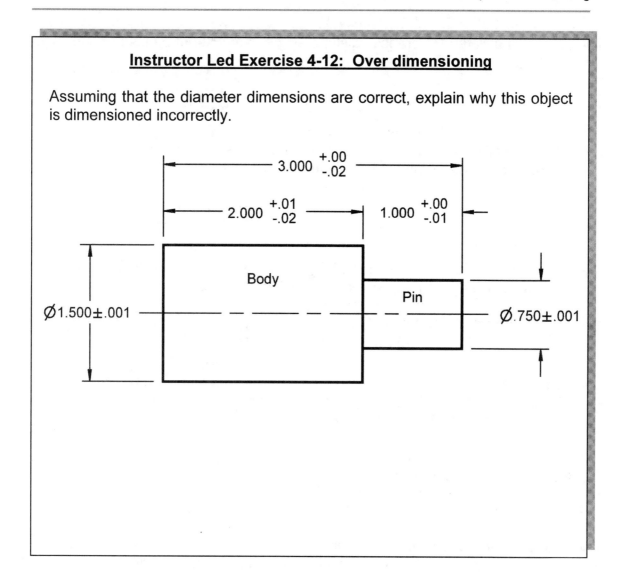

4.9) FORMATTING TOLERANCES

The conventions that are presented in this section pertain to the number of decimal places and format, and are in accordance with the ASME Y14.5M standard.

4.9.1) Metric Tolerances

- If a tolerance is obtained from a standardized fit table, the limits plus the basic size and tolerance symbol should be given in one of the following three ways. It is preferred to use the forms that directly state the limits.

$$\varnothing \, ^{20.240}_{20.110} \, (\varnothing 20 \text{ C11}) \qquad \text{or} \qquad \varnothing 20 \text{ C11} \left(\varnothing \, ^{20.240}_{20.110} \right) \qquad \text{or} \qquad \varnothing 20 \text{ C11}$$

- Where a unilateral tolerance exists, a single zero without a plus or minus sign is shown.

$$40 \,{}^{0}_{-0.02} \qquad \text{or} \qquad 40 \,{}^{+0.02}_{0}$$

- Where a bilateral tolerance is used, both the plus and minus values have the same number of decimal places, using zeros where necessary.

$$10 \,{}^{+0.25}_{-0.10} \qquad \text{not} \qquad \cancel{10 \,{}^{+0.25}_{-0.1}}$$

- If limit dimensions are used, both values should have the same number of decimal places, using zeros where necessary.

$$\begin{matrix} 15.45 \\ 15.00 \end{matrix} \qquad \text{not} \qquad \cancel{\begin{matrix} 15.45 \\ 15 \end{matrix}}$$

- Basic dimensions are considered absolute. When used with a tolerance, the number of decimal places in the basic dimension does not have to match the number of decimal places in the tolerance.

$$45 \pm 0.15 \qquad \text{not} \qquad \cancel{45.00 \pm 0.15}$$

4.9.2) Inch Tolerances

- For unilateral and bilateral tolerances, the basic dimension and the plus and minus values should be expressed with the same number of decimal places.

$$.500 \,{}^{+.000}_{-.002} \qquad \text{not} \qquad \cancel{.500 \,{}^{0}_{-.002}}$$

$$.500 \,{}^{+.001}_{-.002} \qquad \text{not} \qquad \cancel{.50 \,{}^{+.001}_{-.002}}$$

- If limit dimensions are used, both values should have the same number of decimal places, using zeros where necessary.

$$\begin{matrix} .250 \\ .252 \end{matrix} \qquad \text{not} \qquad \cancel{\begin{matrix} .25 \\ .252 \end{matrix}}$$

- When basic dimensions are used, the number of decimal places should match the number of decimal places in the tolerance.

$$2.000\pm0.015 \qquad not \qquad 2.0\pm0.015$$

4.9.3) Angular Tolerances

- Where angle dimensions are used, both the angle and the plus and minus values have the same number of decimal places.

$$30.0°\pm.2° \qquad not \qquad 30°\pm.2°$$

NOTES:

TOLERANCING REVIEW QUESTIONS

Name: _____   Date: _____

Answer the following questions.

Q4-1) Is it possible to machine a part to an exact size?

Q4-2) What is a tolerance?

Q4-3) Why are tolerances necessary?

Q4-4) Name three factors that influence tolerance choice.

Q4-5) A limit dimension for an internal feature is generally given with the (smaller, larger) dimension first or on top.

Q4-6) A limit dimension for an external feature is generally given with the (smaller, larger) dimension first or on top.

Q4-7) What is one advantage for each of the following tolerance types:
        a. Limit dimensions.
        b. Plus or minus tolerances.
        c. Page or block tolerances.

Q4-8) Explain the difference between a clearance fit, an interference fit, and a transition fit.

Q4-9) What do the letter symbols RC, LC, LT, LN, and FN signify?

Q4-10) When designating a metric fit, what does the letter represent and what does the number represent?

Q4-11) When specifying the tolerance zone, does the letter case matter? Why or why not?

Q4-12) What are the two systems used in the metric tolerance tables?

Q4-13) Decreasing a tolerance will (decrease, increase) manufacturing costs.

TOLERANCING PROBLEMS

Name: _____   Date: _____

P4-1) Fill in the given table for the following shaft and hole limits.

	(a)	(b)	(c)	(d)	(e)	(f)	(g)	(h)
Shaft	.9975	4.7494	.494	5.0005	1.2513	.1256	2.5072	8.9980
Limits	.9963	4.7487	.487	4.9995	1.2507	.1254	2.5060	8.9878
Hole	1.000	4.750	.500	5.0000	1.250	.1250	2.5000	9.0000
Limits	1.002	4.751	.507	5.0016	1.251	.1253	2.5018	9.0018

	Shaft	Hole
Limits		
Basic Size		
Tolerance		
MMC		
LMC		
Max. Clearance		
Min. Clearance (Allowance)		
Type of Fit		

	Shaft	Hole
Limits		
Basic Size		
Tolerance		
MMC		
LMC		
Max. Clearance		
Min. Clearance (Allowance)		
Type of Fit		

	Shaft	Hole
Limits		
Basic Size		
Tolerance		
MMC		
LMC		
Max. Clearance		
Min. Clearance (Allowance)		
Type of Fit		

<u>NOTES:</u>

Name: _____   Date: _____

P4-2)   Find the limits of the shaft and hole for the following basic size – fit combinations.

	(a)	(b)	(c)	(d)	(e)	(f)	(g)
Basic Size	.25	.5	.75	1	1.375	2	2.125
Fit 1	RC3	RC5	RC1	RC9	RC7	RC4	RC6
Fit 2	LC2	LC7	LC11	LC6	LC1	LC9	LC4
Fit 3	LT1	LT2	LT3	LT4	LT5	LT6	LT3
Fit 4	LN2	FN3	FN1	FN5	LN3	FN4	FN2

NOTES:

Name: _____ Date: _____

P4-3) Fill in the given table for the following shaft and hole limits.

	(a)	(b)	(c)	(d)	(e)	(f)	(g)	(h)
Shaft	5.970	79.990	16.029	25.061	120.000	2.000	30.000	8.000
Limits	5.940	79.971	16.018	25.048	119.780	1.994	29.987	7.991
Hole	6.000	80.000	16.000	25.000	120.180	2.002	29.985	7.963
Limits	6.030	80.030	16.018	25.021	120.400	2.012	30.006	7.978

	Shaft	Hole
Limits		
Basic size		
Tolerance		
Upper Deviation		
Lower Deviation		
Fundamental Deviation		
IT Grade		
System (hole, shaft)		
Fit		

	Shaft	Hole
Limits		
Basic size		
Tolerance		
Upper Deviation		
Lower Deviation		
Fundamental Deviation		
IT Grade		
System (hole, shaft)		
Fit		

	Shaft	Hole
Limits		
Basic size		
Tolerance		
Upper Deviation		
Lower Deviation		
Fundamental Deviation		
IT Grade		
System (hole, shaft)		
Fit		

NOTES:

Name: _____   Date: _____

P4-4) Find the limits of the shaft and hole for the following basic size – fit combinations.

	(a)	(b)	(c)	(d)	(e)	(f)	(g)
Basic Size	5	10	12	16	20	25	30
Fit 1	H11/c11	H7/k6	H7/p6	H7/u6	H8/f7	H7/n6	H7/s6
Fit 2	U7/h6	N7/h6	G7/h6	C11/h11	S7/h6	H7/h6	K7/h6

NOTES:

P4-5) Draw and dimension the following objects using proper dimensioning techniques. Notice that the dimensioned isometric drawing does not always use the correct symbols or dimensioning techniques.

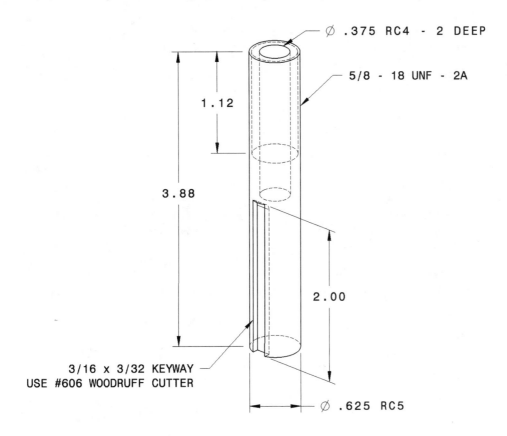

P4-6) Draw and dimension the following objects using proper dimensioning techniques. Notice that the dimensioned isometric drawing does not always use the correct symbols or dimensioning techniques.

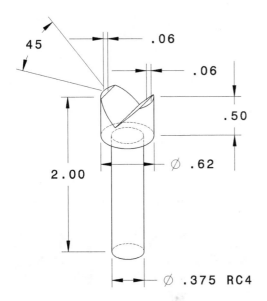

NOTES:

THREADS AND FASTENERS

In Chapter 5 you will learn about fasteners. Fasteners give us the means to assemble parts and to later disassemble them if necessary. Most fasteners have threads; therefore, it is important to understand thread notation when learning about fasteners. By the end of this chapter, you will be able to draw and correctly annotate threads on an orthographic projection. You will also be able to calculate an appropriate bolt or screw clearance hole.

5.1) FASTENERS

Fasteners include items such as bolts, nuts, set screws, washers, keys, and pins, just to name a few. Fasteners are not a permanent means of assembly, such as, welding or adhesives. They are used in the assembly of machines that, in the future, may need to be taken apart and serviced. The most common type of fastener is the screw. There are many types of screws and many types of screw threads or thread forms.

Fasteners and threaded features must be specified on your engineering drawing. Most fasteners are purchased; therefore, specifications must be given to allow the fastener to be ordered correctly.

5.2) SCREW THREAD DEFINITIONS

- <u>Screw Thread:</u> A screw thread is a ridge of uniform section in the form of a helix (see Figure 5-1).

- <u>External Thread:</u> External threads are on the outside of a member. A chamfer on the end of the screw thread makes it easier to engage the nut. An external thread is cut using a die or lathe.

- <u>Internal Thread:</u> Internal threads are on the inside of a member. An internal thread is cut using a tap.

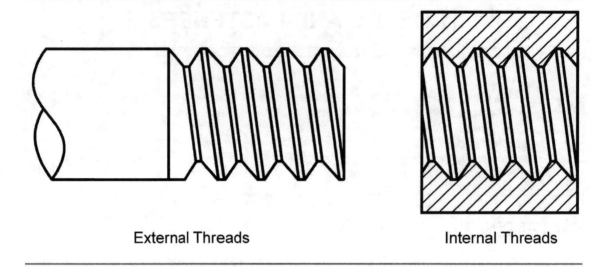

External Threads | Internal Threads

Figure 5-1: External and internal threads.

- Major DIA (D): The major diameter is the largest diameter for both internal and external threads. Sometimes referred to as the nominal or basic size.

- Minor DIA (d): The minor diameter is the smallest diameter. The minor diameter is also referred to as the pitch diameter.

- Crest: The crest is the top surface of the thread.

- Root: The root is the bottom surface of the thread.

- Side: The side is the surface between the crest and root.

- Depth of thread: The depth of thread is the perpendicular distance between the crest and the root and is equal to (D-d)/2.

- Pitch (P): The pitch is the distance from a point on one thread to the corresponding point on the next thread. The pitch is given in inches per threads or millimeters per thread.

- Angle of Thread (A): The angle of thread is the angle between the sides of the threads.

- Screw Axis: The screw axis is the longitudinal centerline.

- Lead: The lead is the distance a screw thread advances axially in one turn.

- <u>Right Handed Thread:</u> Right handed threads advance when turned clockwise (CW). Threads are assumed RH unless specified otherwise.

- <u>Left Handed Thread:</u> Left handed threads advance when turned counter clockwise (CCW).

Application Question 5-1

Name an example of a left handed thread.

Instructor Led Exercise 5-1: Screw thread features

Identify the screw thread features using the preceding definitions as a guide.

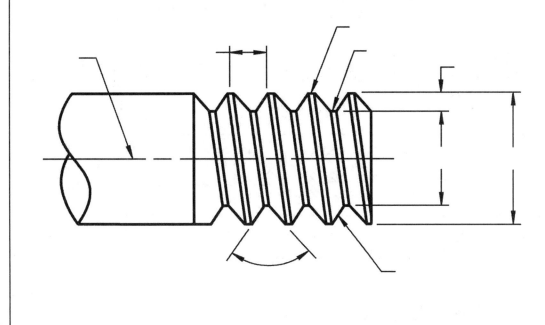

5.3) <u>TYPES OF THREAD</u>

There are many different types of threads or thread forms available. The thread form is the shape of the thread. The choice of thread form that should be used for a particular application depends on length of engagement and load among other factors. *Unified* and *Metric* threads are the most widely used thread forms. Table 5-1 describes just a few of the different thread forms available and their uses.

Thread Name	Figure	Uses
Unified screw thread		General use.
Metric screw thread		General use.
Square		Ideal thread for power transmission.
ACME		Stronger than square thread.
Buttress		Designed to handle heavy forces in one direction (e.g. truck jack).

Table 5-1: Screw thread examples

5.4) <u>MANUFACTURING SCREW THREADS</u>

Before proceeding with a description of how to draw screw threads, it is helpful to understand the manufacturing processes used to produce threads.

To cut internal threads, a tap drill hole is drilled first and then the threads are cut using a tap. The tap drill hole is a little bigger than the minor diameter of the mating external thread to allow engagement. The depth of the tap drill is longer than the length of the threads to allow the proper amount of threads to be cut as shown in Figure 5-2. There are approximately three useless threads at the end of a normal tap. A bottom tap has useful threads all the way to the end, but is more expensive than a normal tap. If a bottom tap is used, the tap drill depth is the same as the thread length.

To cut external threads, you start with a shaft the same size as the major diameter. Then, the threads are cut using a die or on a lathe. For both internal and external threads, a chamfer is usually cut at the points of engagement to allow easy assembly.

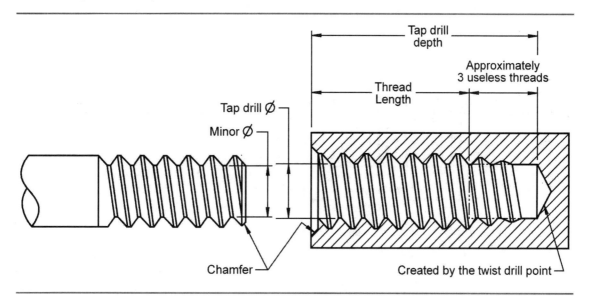

Figure 5-2: Manufacturing screw threads.

5.5) <u>DRAWING SCREW THREADS</u>

There are three methods of representing screw threads on a drawing: *detailed*, *schematic*, and *simplified*. The screw thread representations and standards presented in this chapter are in accordance with the ASME Y14.6-2001 standard.

5.5.1) <u>Detailed Representation</u>

A detailed representation is a close approximation of the appearance of an actual screw thread. The form of the thread is simplified by showing the helix structure with straight lines and the truncated crests and roots as a sharp 'V' similar to that shown in Figure 5-1. This method is comparatively difficult and time consuming.

5.5.2) <u>Schematic Representation</u>

The schematic representation is nearly as effective as the detailed representation and is much easier to draw. Staggered lines are used to represent the thread roots and crests (see Figures 5-3 and 5-4). This method should not be used for hidden internal threads or sections of external threads.

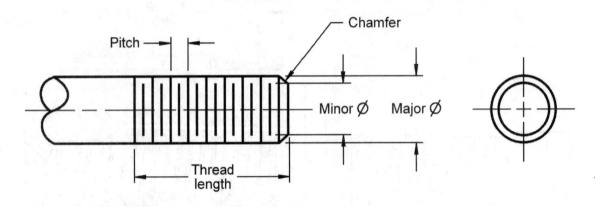

Figure 5-3: Schematic representation of external threads.

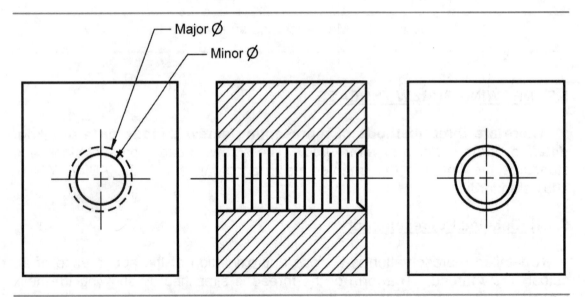

Figure 5-4: Schematic representation of internal threads.

5.5.3) <u>Simplified Representation</u>

In the simplified representation, the screw threads are drawn using visible and hidden lines to represent the major and minor diameters. Line choice depends on whether the thread is internal or external and the viewing direction (see Figures 5-5 and 5-6). Simplified threads are the simplest and fastest to draw. This method should be used whenever possible.

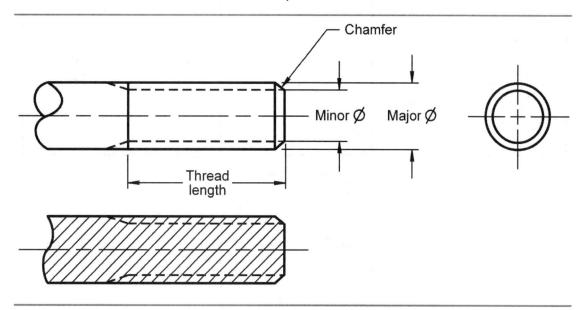

Figure 5-5: Simplified representation of external threads.

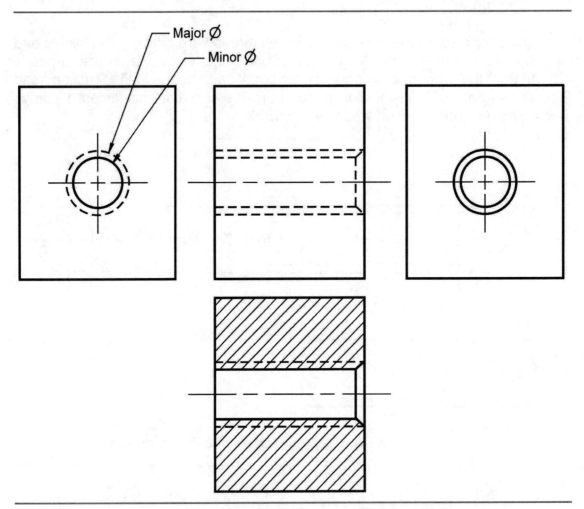

Figure 5-6a: Simplified representation of internal threads.

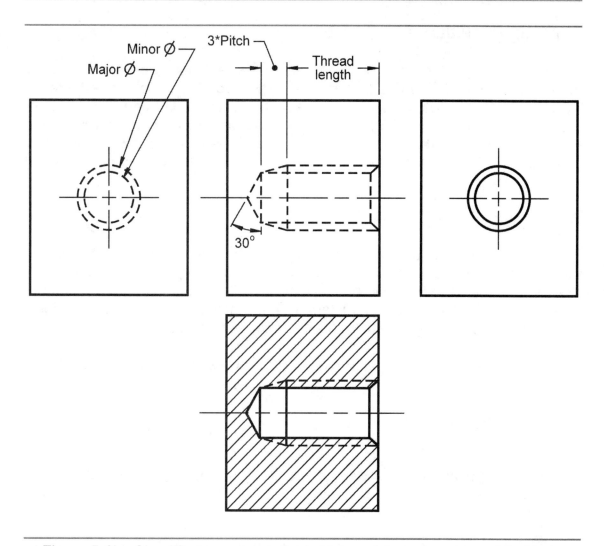

Figure 5-6b: Simplified representation of internal threads cut on a blind hole.

If screw thread tables are not available for reference, the minor diameter can be approximated as 75% of the major diameter.

5.6) UNIFIED THREADS

After drawing a thread using the proper representation, we need to identify the thread form and type in a thread note. Each type of thread (Unified, Metric, ACME, etc...) has its own way of being identified. **Unified threads are identified in a thread note by their *major diameter*, *threads per inch*, *thread form* and *series*, *thread class*, whether the thread is *external* or *internal*, whether the thread is *right* or *left* handed, and the thread *depth* (internal only).**

5.6.1) Unified Thread Note

The following is a list of components that should be included in the Unified thread note. The first three components (major diameter, threads per inch, and thread form and series) should be included in all thread notes and the depth of thread should be included for all internal thread notes. The other components are optional and are only used if additional refinement is needed.

1. Major Diameter: The major diameter is the largest diameter for both internal and external threads.

2. Threads per Inch: The number of threads per inch is equal to one over the pitch.

3. Thread Form and Series: The thread form is the shape of the thread cut (Unified) and the thread series is the number of threads per inch for a particular diameter (coarse, fine, extra fine).

   - UNC: UNC stands for *Unified National coarse*. Coarse threads are the most commonly used thread.

   - UNF: UNF stands for *Unified National fine*. Fine threads are used when high degree of tightness is required.

   - UNEF: UNEF stands for *Unified National extra fine*. Extra fine threads are used when the length of engagement is limited (e.g. sheet metal).

4. Thread Class: The thread class indicates the closeness of fit between the two mating threaded parts. There are three thread classes. A thread class of "1" indicates a generous tolerance and used when rapid assembly and disassembly is required. A thread class of "2" is a normal production fit. This fit is assumed if none is stated. A thread class of "3" is used when high accuracy is required.

5. <u>External or Internal Threads:</u> An "A" (external threads) or "B" (internal threads) is placed next to the thread class to indicate whether the threads are external or internal.

6. <u>Right handed or left handed thread:</u> Right handed threads are indicated by the symbol "RH" and left handed threads are indicated by the symbol "LH". Right handed threads are assumed if none is stated.

7. <u>Depth of thread:</u> The thread depth is given at the end of the thread note and indicates the thread depth for internal threads. The stated depth is not the tap drill depth. Remember the tap drill depth is longer that the thread depth.

Instructor Led Exercise 5-2: Unified National thread note components

Identify the different components of the following Unified National thread notes.

1/4 – 20 UNC – 2A – RH

1/4	
20	
UNC	
2	
A	
RH	

1/4 – 28 UNF – 3B – LH

1/4	
28	
UNF	
3	
B	
LH	

5.6.2) Unified Thread Tables

Standard screw thread tables are available in order to look up the major diameter, threads per inch, tap drill size, and minor diameter for a particular thread. These thread tables are given in the ASME B1.1-2003 standard which are given in Appendix E.

Instructor Led Exercise 5-3: Unified National thread note

Write the thread note for a #10 fine thread.

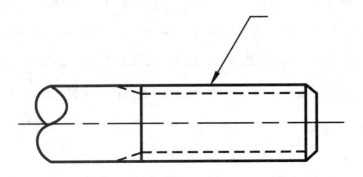

What are the major and minor diameters in inches?

Major DIA	
Minor DIA	

5.7) METRIC THREADS

Metric threads are identified, in a thread note, by "M" for Metric thread form, the *major diameter* followed by a lower case "x", *pitch*, *tolerance class*, whether the thread is *right* or *left* handed, and *thread depth* (internal only).

5.7.1) Metric thread note

The following is a list of components that should be included in a Metric thread note. The first three components (Metric form, major diameter, and pitch) should be included in all thread notes and the depth of thread should be included for all internal thread notes. The other components are optional and are only used if additional refinement is needed.

1. Metric Form: Placing an "M" before the major diameter indicates the Metric thread form.

2. Major Diameter: This major diameter is the largest diameter for both internal and external.

3. Pitch: The pitch is given in millimeters per thread.

4. Tolerance Class: The tolerance class describes the looseness or tightness of fit between the internal and external threads. The tolerance class contains both a tolerance grade given by a number and tolerance position given by a letter. In a thread note, the minor or pitch diameter tolerance is stated first followed by the major or crest diameter tolerance if it is different. Two classes of Metric thread fits are generally recognized. For general purpose, the fit "6H/6g" should be used. This fit is assumed if none is stated. For a closer fit, use "6H/5g6g".

   - Tolerance Grade: The tolerance grade is indicated by a number. The smaller the number the tighter the fit. The number "5" indicates good commercial practice. The number "6" is for general purpose threads and is equivalent to the thread class "2" used for Unified National threads.

   - Tolerance Position: The tolerance position specifies the amount of allowance and is indicated by a letter. Upper case letters are used for internal threads and lower case letters for external threads. The letter "e" is used for large allowances, "g" and "G" are used for small allowances, and "h" and "H" are used for no allowance.

5. <u>Right handed or left handed thread:</u> Right handed threads are indicated by the symbol "RH" and left handed threads are indicated by the symbol "LH". Right handed threads are assumed if none is stated.

6. <u>Depth of thread:</u> The thread depth is given at the end of the thread note and indicates the thread depth for internal threads, not the tap drill depth.

Instructor Led Exercise 5-4: Metric thread note components

Identify the different components of the following Metric thread notes.

M10x1.5 – 4h6h – RH

M	
10	
1.5	
4h	
6h	
Internal or External	
RH	

M10x1.25 – 5H6H – LH

M	
10	
1.25	
5H	
6H	
Internal or External	
LH	

5.7.2) Metric Thread Tables

Standard screw thread tables are available in order to look up the major diameter, threads per inch, tap drill size, and minor diameter for a particular thread. These thread tables are given in the ASME B1.13M-2001 standard which are given in Appendix E.

Instructor Led Exercise 5-5: Metric thread tables

For a ∅16 internal Metric thread, what are the two available pitches and the corresponding tap drill diameter and the corresponding minor diameter for the mating external thread?

Pitch	Tap drill size	Minor DIA

Which has the finer thread?

The finer thread is M16x(   )

Write the thread note for a 16 mm diameter coarse thread.

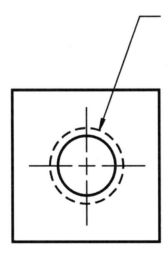

5.8) **DRAWING BOLTS**

Figure 5-7 illustrates how to draw bolts. The variable D represents the major or nominal diameter of the bolt. Nuts are drawn in a similar fashion.

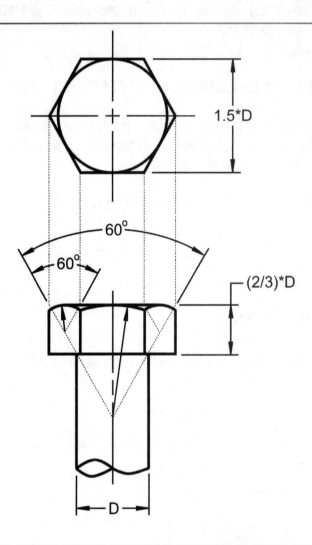

Figure 5-7: Drawing bolts.

5.9) <u>BOLT AND SCREW CLEARANCES</u>

Bolts and screws attach one material with a clearance hole to another material with a threaded hole. The size of the clearance hole depends on the major diameter of the fastener and the type of fit that is required for the assembly to function properly. Clearance holes can be designed to have a *normal*, *close* or *loose* fit. Table 5-2 gives the normal fit clearances which are illustrated in Figure 5-8. For detailed information on clearances for bolts and screws, refer to the ASME B18.2.8-1999 standard also given in Appendix E.

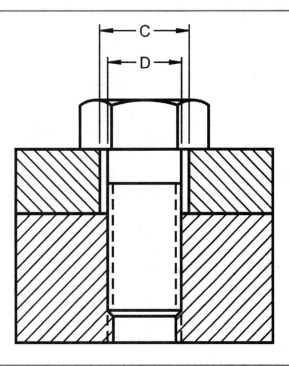

Figure 5-8: Bolt clearance.

Inch clearances	
Nominal screw size (D)	Clearance hole (C)
#0 - #4	D + 1/64
#5 – 7/16	D + 1/32
1/2 – 7/8	D + 1/16
1	D + 3/92
1 1/8, 1 1/4	D + 3/32
1 3/8, 1 1/2	D + 1/8

Metric clearances	
Nominal screw size (D)	Clearance hole (C)
M1.6	D + 0.2
M2, M2.5	D + 0.4
M4, M5	D + 0.5
M6	D + 0.6
M8, M10	D + 1
M12 – M16	D + 1.5
M20, M24	D + 2
M30 – M42	D + 3
M48	D + 4
M56 – M90	D + 6
M100	D + 7

Table 5-2: Bolt and screw normal fit clearance holes.

Sometimes bolt or screw heads need to be flush with the surface. This can be achieved by using either a counterbore or countersink depending on the head shape. Counterbores are holes that are designed to recess bolt or screw heads below the surface of a part as shown in Figure 5-9. Countersinks are angled holes that are designed to recess screws with angled heads as shown in Figure 5-10. Appendix E gives the clearance hole diameters illustrated in Figure 5-9 and 5-10. Typically CH = H + 1/16 (1.5 mm) and C1 = D + 1/8 (3 mm).

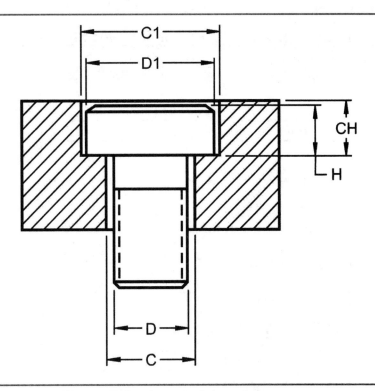

Figure 5-9: Counterbore clearances.

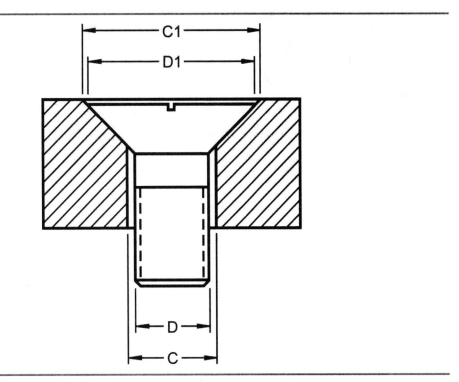

Figure 5-10: Countersink clearances.

Instructor Led Exercise 5-6: Fastener tables and clearance holes

What is the normal fit clearance hole diameter for the following nominal bolt sizes?

Nominal size	
1/4	
3/4	

A 5/16 - 18 UNC – Socket Head Cap Screw needs to go through a piece of metal in order to screw into a plate below. The head of the screw should be flush with the surface. Fill in the following table for a normal fit clearance hole. Refer to Appendix E.

Head diameter	
Height of head	
Clearance hole diameter	
Counterbore diameter	
Counterbore depth	

An M8x1.25 Slotted Flat Countersunk Head Metric Machine Screw needs to go through a piece of metal in order to screw into a plate below. The clearance hole needs to be close and the head needs to be flush with the surface. What should the countersink diameter and clearance hole diameter be?

Major diameter	
Head diameter	
Countersink diameter	
Clearance hole diameter	

THREAD AND FASTENER REVIEW QUESTIONS

Name: _____ Date: _____

Answer the following questions.

Q5-1) What are the units of pitch?

Q5-2) The tap drill size is closest in size to the (major diameter, minor diameter).

Q5-3) Explain the difference between the schematic and simplified thread symbols.

Q5-4) How much longer is the tap drill depth than the thread depth when using a normal tap to cut the threads?

Q5-5) For Unified National threads, explain the difference between the thread series and thread form.

Q5-6) In the Unified National thread note, we specify the (pitch, 1/pitch).

Q5-7) In the Metric form thread note, we specify the (pitch, 1/pitch).

Q5-8) Upon what does a bolt or screw clearance hole diameter depend?

<u>NOTES:</u>

THREAD AND FASTENER PROBLEMS

Name: _____   Date: _____

P5-1) Write the thread notes for the following external threads. Also, what are the minor diameter and the pitch? Thread class = 2.

	(a)	(b)	(c)	(d)	(e)	(f)	(g)
Major ⌀	1/4	7/8	3/4	1/2	1	3/8	5/16
Series	Fine	Coarse	Fine	Coarse	Fine	Extra Fine	Coarse

P5-2) Write the thread notes for the following internal threads. Also, what are the tap drill size and/or diameter and the pitch? Thread class = 3.

	(a)	(b)	(c)	(d)	(e)	(f)	(g)
Major ⌀	7/16	1/4	5/8	1 ¼	3/8	1/2	1
Series	Fine	Coarse	Fine	Coarse	Fine	Extra Fine	Coarse

P5-3) Write the thread notes for the following threads. Also, what is the major diameter in inches?

	(a)	(b)	(c)	(d)	(e)	(f)	(g)
Major ⌀	#0	#2	#4	#5	#6	#8	#10
Series	Fine	Coarse	Fine	Coarse	Fine	Extra Fine	Coarse

NOTES:

Name: _____   Date: _____

P5-4) Write the thread notes for the following external threads. Also, what are the minor diameter and the number of threads per mm?

	(a)	(b)	(c)	(d)	(e)	(f)	(g)
Major ⌀	M3	M4	M8	M10	M12	M20	M24
Series	Fine	Coarse	Fine	Coarse	Fine	Fine	Coarse

P5-5) Write the thread notes for the following internal threads. Also, what are the tap drill size and/or diameter and the number of threads per mm?

	(a)	(b)	(c)	(d)	(e)	(f)	(g)
Major ⌀	M1.6	M5	M6	M12	M18	M22	M27
Series	Fine	Coarse	Fine	Coarse	Fine	Fine	Coarse

NOTES:

Name: _____   Date: _____

P5-6) Fill in the given table for a hex head bolt with the following major diameters.

	(a)	(b)	(c)	(d)	(e)	(f)	(g)	(h)
Major ⌀	1/4	5/16	1/2	7/8	1	9/16	3/8	7/16

Max. width across flats	
Max. width across corners	
Max. head height	
Wrenching height	
Thread length for a screw that is shorter than 6 inches	
Normal clearance hole	

Max. width across flats	
Max. width across corners	
Max. head height	
Wrenching height	
Thread length for a screw that is shorter than 6 inches	
Normal clearance hole	

Max. width across flats	
Max. width across corners	
Max. head height	
Wrenching height	
Thread length for a screw that is shorter than 6 inches	
Normal clearance hole	

Max. width across flats	
Max. width across corners	
Max. head height	
Wrenching height	
Thread length for a screw that is shorter than 6 inches	
Normal clearance hole	

NOTES:

Name: _____   Date: _____

P5-7) Fill in the given table for a hexagon (socket) head cap screw with the following major diameters.

	(a)	(b)	(c)	(d)	(e)	(f)	(g)	(h)
Major ⌀	1/4	5/16	1/2	#8	#5	9/16	3/8	#10

Max. head diameter	
Max. head height	
Normal clearance hole	
Counterbore diameter	
Counterbore depth	

Max. head diameter	
Max. head height	
Normal clearance hole	
Counterbore diameter	
Counterbore depth	

Max. head diameter	
Max. head height	
Normal clearance hole	
Counterbore diameter	
Counterbore depth	

Max. head diameter	
Max. head height	
Normal clearance hole	
Counterbore diameter	
Counterbore depth	

Max. head diameter	
Max. head height	
Normal clearance hole	
Counterbore diameter	
Counterbore depth	

NOTES:

Name: _____ Date: _____

5-8) Fill in the given table for a slotted flat countersunk head cap screw with the following major diameters.

	(a)	(b)	(c)	(d)	(e)	(f)	(g)	(h)
Major ⌀	1/4	5/16	3/8	7/16	1/2	9/16	5/8	3/4

Max. head diameter	
Max. head height	
Normal clearance hole	
Countersink diameter	
Countersink angle	

Max. head diameter	
Max. head height	
Normal clearance hole	
Countersink diameter	
Countersink angle	

Max. head diameter	
Max. head height	
Normal clearance hole	
Countersink diameter	
Countersink angle	

Max. head diameter	
Max. head height	
Normal clearance hole	
Countersink diameter	
Countersink angle	

Max. head diameter	
Max. head height	
Normal clearance hole	
Countersink diameter	
Countersink angle	

NOTES:

Name: _____ Date: _____

P5-9) Fill in the given table for a hex head bolt with the following major diameters.

	(a)	(b)	(c)	(d)	(e)	(f)	(g)	(h)
Major ⌀	M5	M12	M20	M30	M36	M48	M14	M24

Max. width across flats	
Max. width across corners	
Max. head height	
Wrenching height	
Thread length for a screw that is shorter than 125 mm	
Normal clearance hole	

Max. width across flats	
Max. width across corners	
Max. head height	
Wrenching height	
Thread length for a screw that is shorter than 125 mm	
Normal clearance hole	

Max. width across flats	
Max. width across corners	
Max. head height	
Wrenching height	
Thread length for a screw that is shorter than 125 mm	
Normal clearance hole	

Max. width across flats	
Max. width across corners	
Max. head height	
Wrenching height	
Thread length for a screw that is shorter than 125 mm	
Normal clearance hole	

NOTES:

Name: _____   Date: _____

P5-10) Fill in the given table for a socket head cap screw with the following major diameters.

	(a)	(b)	(c)	(d)	(e)	(f)	(g)	(h)
Major ⌀	M1.6	M2.5	M4	M6	M12	M16	M24	M42

Max. head diameter	
Max. head height	
Normal clearance hole	
Counterbore diameter	
Counterbore depth	

Max. head diameter	
Max. head height	
Normal clearance hole	
Counterbore diameter	
Counterbore depth	

Max. head diameter	
Max. head height	
Normal clearance hole	
Counterbore diameter	
Counterbore depth	

Max. head diameter	
Max. head height	
Normal clearance hole	
Counterbore diameter	
Counterbore depth	

Max. head diameter	
Max. head height	
Normal clearance hole	
Counterbore diameter	
Counterbore depth	

NOTES:

Name: _____   Date: _____

P5-11) Fill in the given table for a slotted flat head machine screw with the following major diameters.

	(a)	(b)	(c)	(d)	(e)	(f)	(g)	(h)
Major ⌀	M2.5	M3	M3.5	M5	M6	M8	M10	M4

Max. head diameter	
Max. head height	
Normal clearance hole	
Countersink diameter	
Countersink angle	

Max. head diameter	
Max. head height	
Normal clearance hole	
Countersink diameter	
Countersink angle	

Max. head diameter	
Max. head height	
Normal clearance hole	
Countersink diameter	
Countersink angle	

Max. head diameter	
Max. head height	
Normal clearance hole	
Countersink diameter	
Countersink angle	

Max. head diameter	
Max. head height	
Normal clearance hole	
Countersink diameter	
Countersink angle	

NOTES:

P5-12) Draw and dimension the following objects using proper dimensioning techniques. Notice that the dimensioning isometric drawing does not always use the proper symbols or dimensioning techniques.

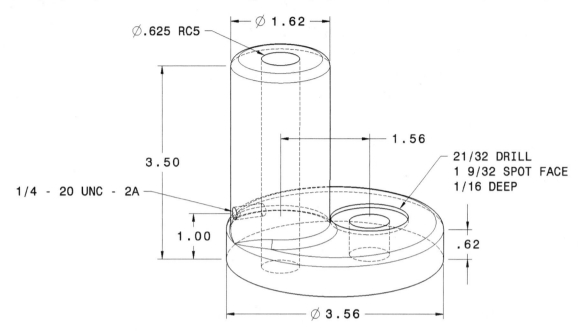

P5-13) Draw and dimension the following objects using proper dimensioning techniques. Notice that the dimensioning isometric drawing does not always use the proper symbols or dimensioning techniques.

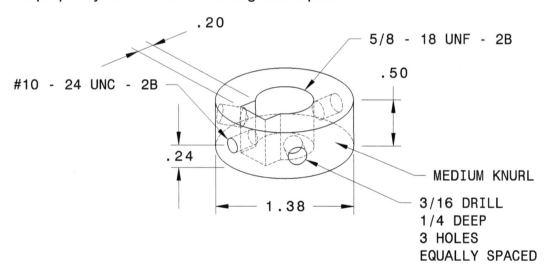

P5-14) Draw and dimension the following objects using proper dimensioning techniques. Notice that the dimensioning isometric drawing does not always use the proper symbols or dimensioning techniques.

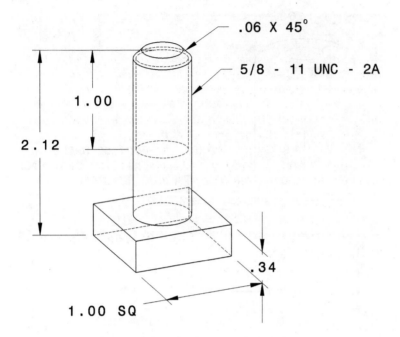

ASSEMBLY DRAWINGS

In Chapter 6 you will learn how to create an assembly drawing. An assembly drawing is a drawing of an entire machine with each part identified. After each part of a machine is manufactured, the assembly drawing shows us how to put these parts together. You will also learn how to generate a standard parts sheet. This sheet contains information about purchased items. The assembly drawing together with all the detailed part drawings and the standard parts sheet is called a working drawing package. By the end of this chapter, you will be able to create a working drawing package which contains all the information necessary to manufacture a machine or system.

An assembly drawing is a drawing of an entire machine or system with all of its components located and identified.

6.1) DEFINITIONS

- Detail Drawing: A detail drawing is a drawing of an individual part, which includes an orthographic projection and dimensions. One detail per sheet.

- Assembly Drawing: An assembly consists of a number of parts that are joined together to perform a specific function (e.g. a bicycle). The assembly may be disassembled without destroying any part of the assembly. An assembly drawing shows the assembled machine or structure with all of the parts in their functional position. Figures 6-1 and 6-2 are examples of assembly drawings.

- Subassembly Drawing: A subassembly is two or more parts that form a portion of an assembly (e.g. the drive train of a bicycle). A subassembly drawing shows only one unit of a larger machine.

- Working Drawing Package: A typical working drawing package includes an assembly drawing, detailed drawings, and a standard parts sheet. The drawing package contains the specifications that will enable the design to be manufactured.

6.1.1) Drawing Order

Drawings included in a working drawing package should be presented in the following order:

1) Assembly drawing (first sheet)
2) Part Number 1
3) Part Number 2
4) ....
5) Standard parts sheet (last sheet)

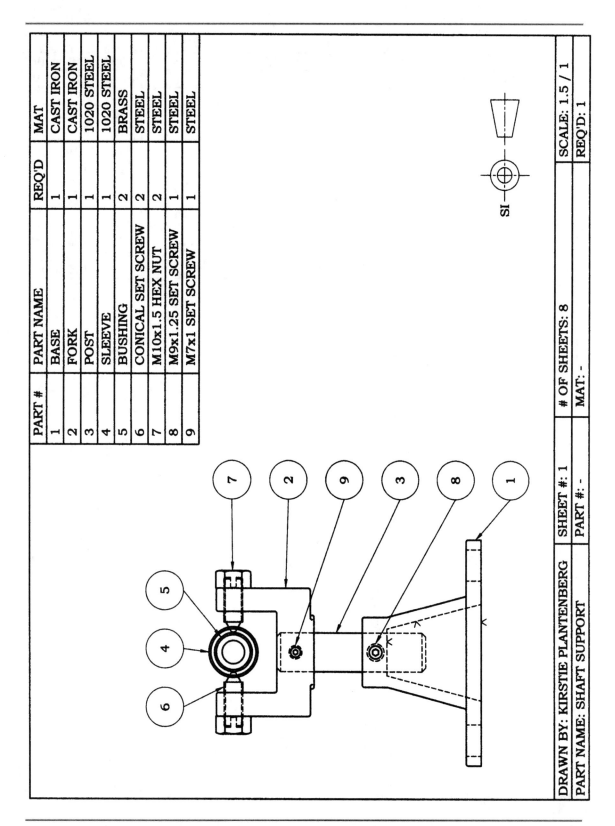

PART #	PART NAME	REQ'D	MAT
1	BASE	1	CAST IRON
2	FORK	1	CAST IRON
3	POST	1	1020 STEEL
4	SLEEVE	1	1020 STEEL
5	BUSHING	2	BRASS
6	CONICAL SET SCREW	2	STEEL
7	M10x1.5 HEX NUT	2	STEEL
8	M9x1.25 SET SCREW	1	STEEL
9	M7x1 SET SCREW	1	STEEL

DRAWN BY: KIRSTIE PLANTENBERG	SHEET #: 1	# OF SHEETS: 8	SCALE: 1.5 / 1
PART NAME: SHAFT SUPPORT	PART #: -	MAT: -	REQ'D: 1

Figure 6-1: Shaft support assembly drawing.

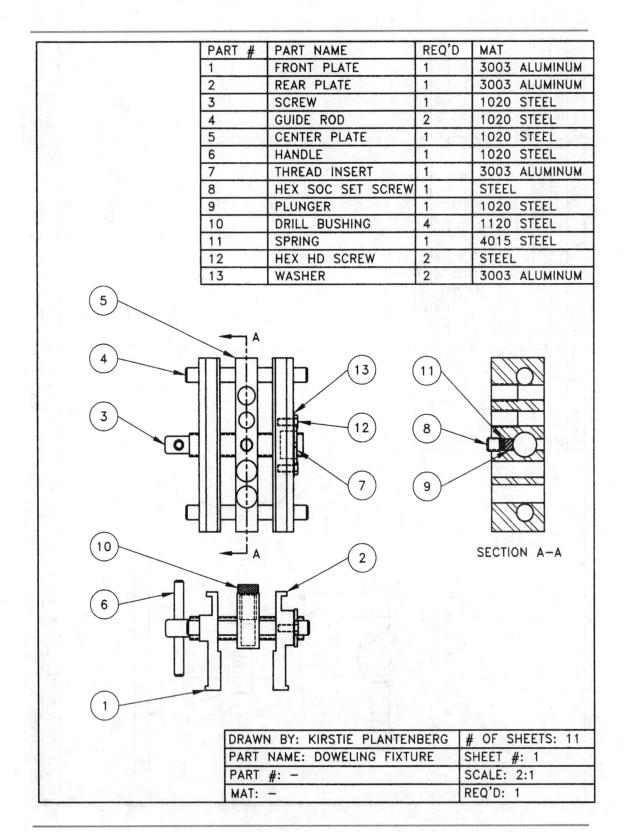

PART #	PART NAME	REQ'D	MAT
1	FRONT PLATE	1	3003 ALUMINUM
2	REAR PLATE	1	3003 ALUMINUM
3	SCREW	1	1020 STEEL
4	GUIDE ROD	2	1020 STEEL
5	CENTER PLATE	1	1020 STEEL
6	HANDLE	1	1020 STEEL
7	THREAD INSERT	1	3003 ALUMINUM
8	HEX SOC SET SCREW	1	STEEL
9	PLUNGER	1	1020 STEEL
10	DRILL BUSHING	4	1120 STEEL
11	SPRING	1	4015 STEEL
12	HEX HD SCREW	2	STEEL
13	WASHER	2	3003 ALUMINUM

SECTION A-A

DRAWN BY: KIRSTIE PLANTENBERG	# OF SHEETS: 11
PART NAME: DOWELING FIXTURE	SHEET #: 1
PART #: −	SCALE: 2:1
MAT: −	REQ'D: 1

Figure 6-2: Doweling fixture assembly drawing.

6.2) VIEWS USED IN ASSEMBLY DRAWINGS

6.2.1) Selecting Views

The purpose of an assembly drawing must be kept in mind when choosing which views need to be included. **The purpose of an assembly drawing is to show how the parts fit together** and to suggest the function of the entire unit. Its purpose is not to describe the shapes of the individual parts. Sometimes only one view is needed and sometimes it is necessary to draw all three principle views. It may also be necessary to include sectional views.

6.2.2) Sectional Views

Since assemblies often have parts fitting into or overlapping other parts, sectioning can be used to great advantage.

- Section Lines: When using sectional views in assembly drawings, it is necessary to distinguish between adjacent parts. **Section lines in adjacent parts are drawn in opposing directions.** In the largest area, the section lines are drawn at 45°. In the next largest area, the section lines are drawn at 135° (in the opposite direction of the largest area). Section line angles of 30° and 60° are used for additional parts. **The distance between the section lines may also be varied to further distinguish between parts**.

Instructor Led Exercise 6-1: Section lines in assemblies

The following assembly is sectioned. Draw in the section lines according to the rules stated above.

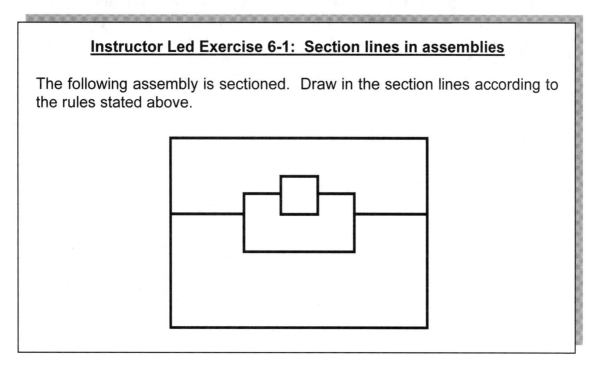

6.3) **THINGS TO INCLUDE/NOT INCLUDE**

The purpose of an assembly drawing is to show how the individual parts fit together. Therefore, each individual part must be identified. It is not; however, used as a manufacturing print. Some lines that were included and necessary in the detailed drawing may be left off the assembly drawing to enhance clearness. The assembly drawing should not look overly cluttered.

6.3.1) **Hidden Lines**

Hidden lines are often not needed. However, they should be used wherever necessary for clearness. It is left to the judgment of the drafter whether or not to include hidden lines. When a section view is used, hidden lines should not be used in the sectional view.

6.3.2) **Dimensions**

As a rule, dimensions are not given on assembly drawings. If dimensions are given, they are limited to some function of the object as a whole.

6.3.3) **Identification**

A part is located and identified by using a circle or balloon containing a part number and a leader line that points to the corresponding part. A balloon containing a part number is placed adjacent to the part. A leader line, starting at the balloon, points to the part to which it refers. Balloons identifying different parts are placed in orderly horizontal or vertical rows. The leader lines are never allowed to cross and adjacent leader lines should be as parallel to each other as possible.

6.3.4) **Parts List/ Bill of Material**

The parts list is an itemized list of the parts that make up the assembled machine. **A parts list contains the *part number*, *part name*, the *number required* and the *material* of the part.** Other information may be included, such as, stock sizes of materials and weights of the parts. Parts are listed in order of their part number. Part numbers are usually assigned based on the size or importance of the part. The parts list is placed either in the upper right corner of the drawing, with part number 1 at the top, or lower right corner of the drawing, with part number 1 at the bottom.

6.4) **STANDARD PARTS**

Standard parts include any part that can be bought off the shelf. **Standard parts do not need to be drawn.** This could include bolts, nuts, washers, keys, etc. Purchasing information is specified on a standard parts sheet attached to the back of a working drawing package. Figure 6-3 shows an example of a standard parts sheet. This standard parts sheet lists four different items. The circle/balloon next to the item contains the part number.

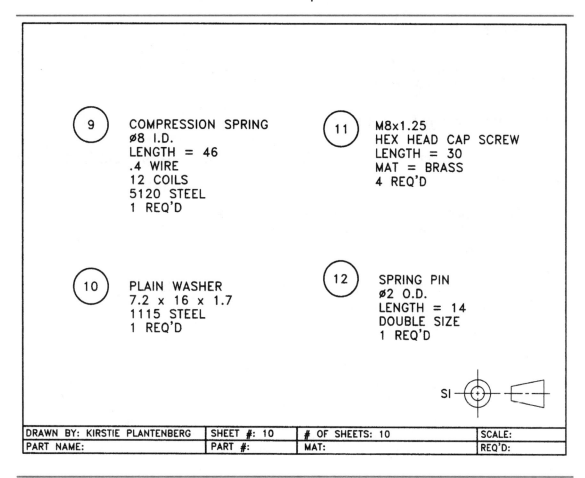

Figure 6-3: Standard parts sheet.

6.4.1) <u>General Fastener Specifications</u>

The information that should be specified on a standard parts sheet for a general fastener is listed below.

1. Thread specification (only if the fastener contains threads)
2. Name of fastener
3. Head/point style or shape (if applicable)
4. Fastener length or size
5. Fastener series
6. Material
7. Special requirements (coatings, finishes, specifications to meet)
8. REQ'D (i.e. number required)

6.4.2) <u>Specifications for Bolts and Nuts</u>

The information that should be specified on a standard parts sheet for a bolt or nut is listed below.

1. Thread specification contained in the thread note
2. Style of head and name of the bolt or nut
3. Length of bolt
4. Material
5. Special requirements (coatings, finishes, specifications to meet)
6. REQ'D (i.e. number required)

In Class Student Exercise 6-2: Working drawing package 1

Consider the *Trolley* assembly shown. Draw detailed drawings of the individual parts, create a standard parts sheet, and draw an assembly drawing. The Trolley dimensions are shown in the exploded assemblies shown on the following pages.

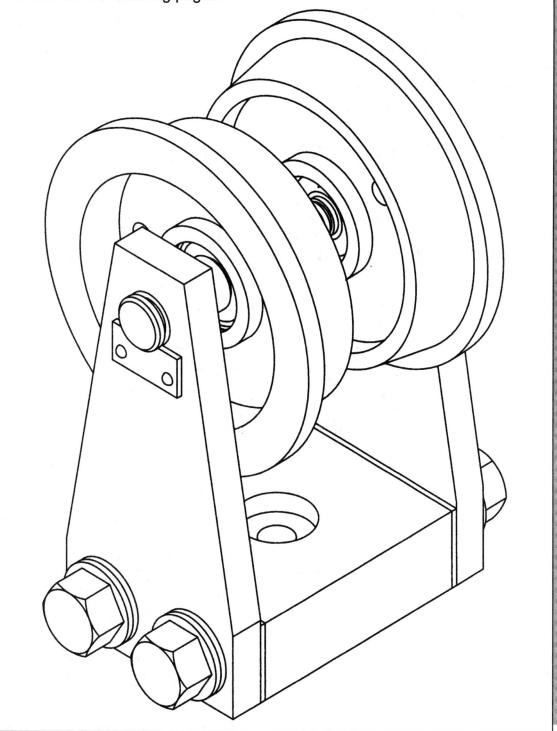

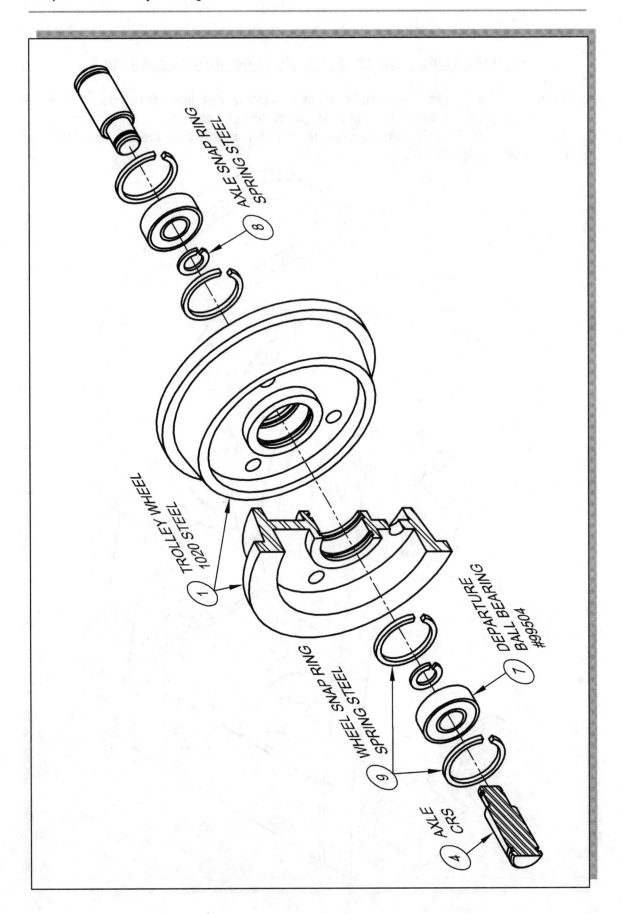

AXLE SNAP RING
SPRING STEEL

⑧

TROLLEY WHEEL
1020 STEEL

①

WHEEL SNAP RING
SPRING STEEL

⑨

DEPARTURE
BALL BEARING
#99504

⑦

AXLE
CRS

④

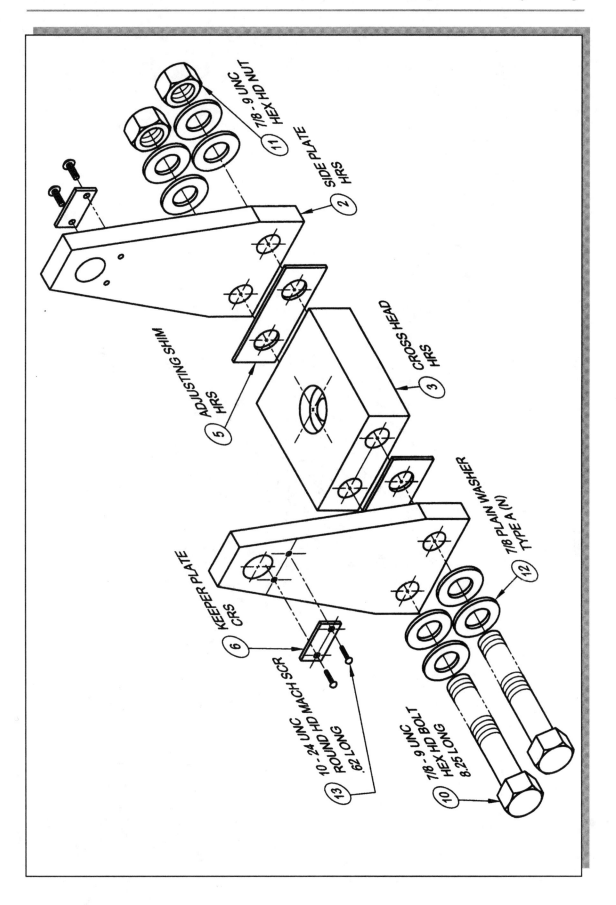

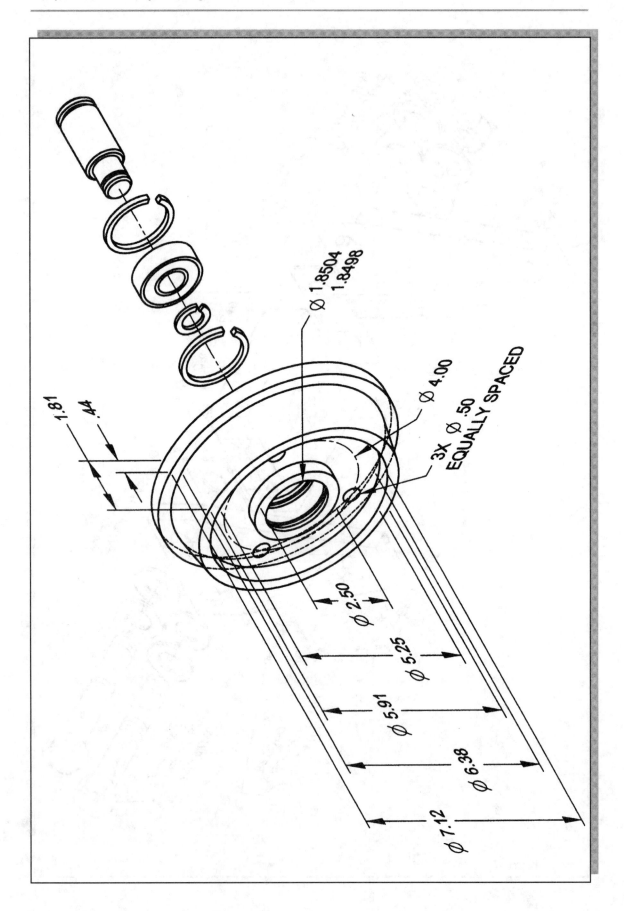

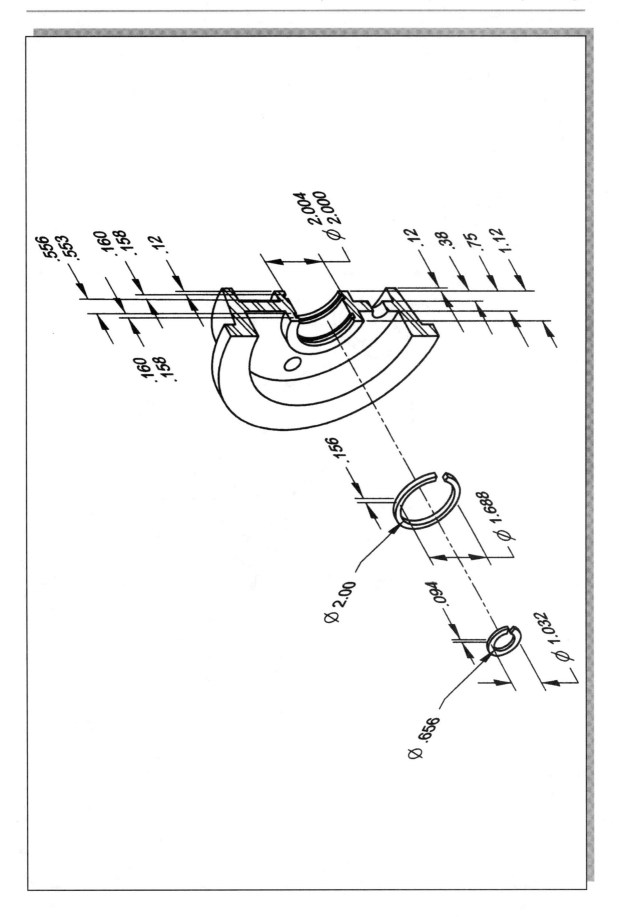

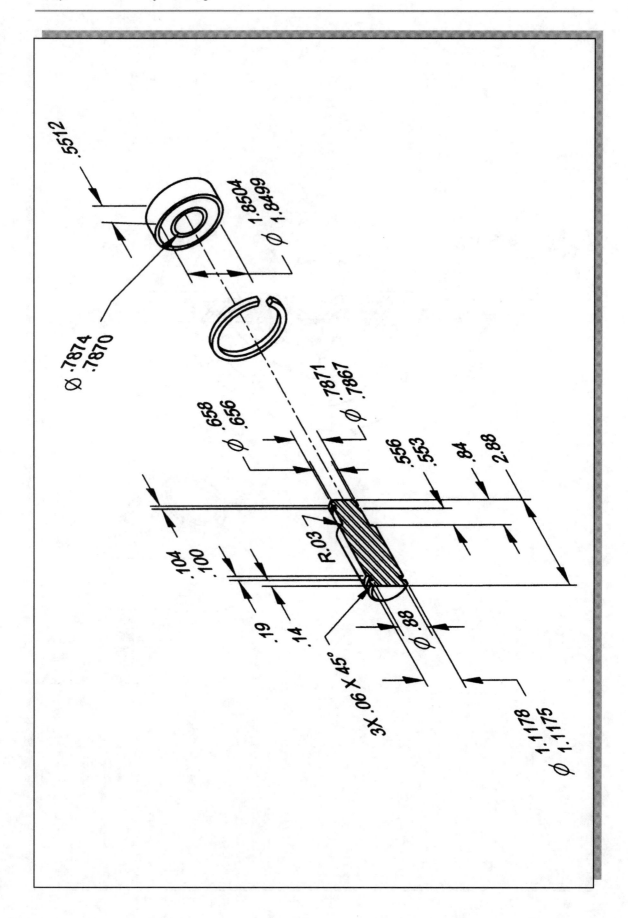

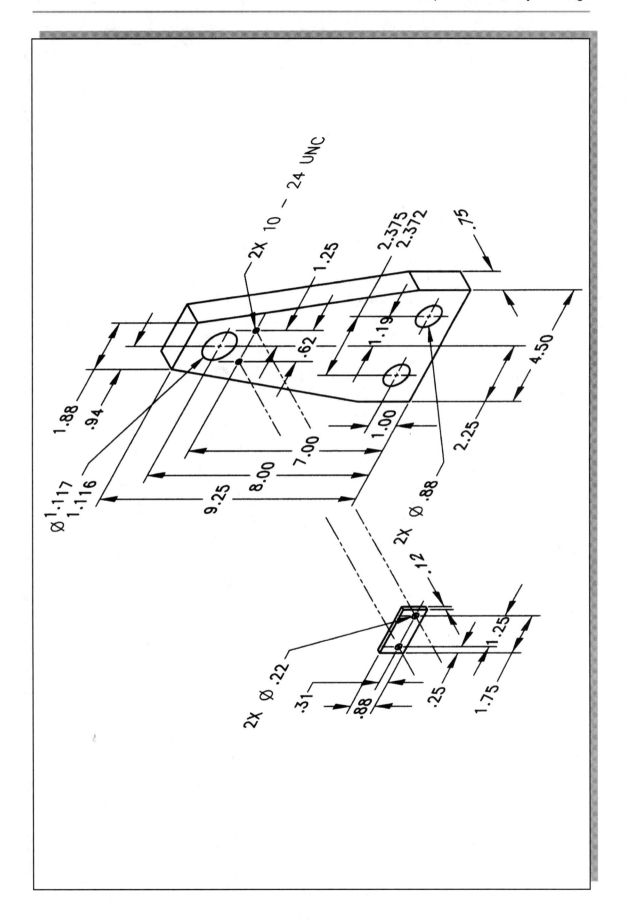

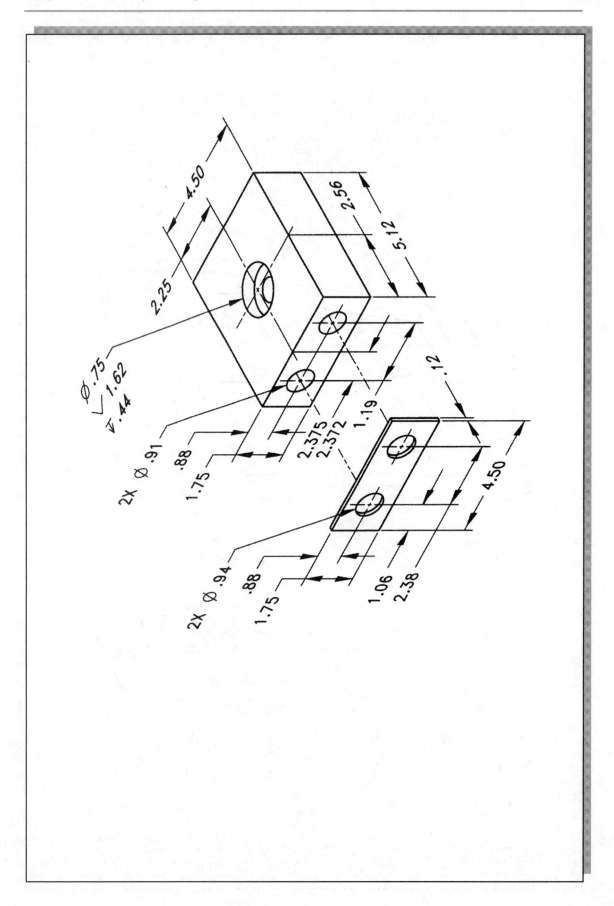

Name: _____   Date: _____

Fill in the part numbers in the appropriate balloons and complete the part list and title block information.

PART #	PART NAME	REQ'D	MAT
1			
2			
3			
4			
5			
6			
7			
8			
9			
10			
11			
12			
13			

DRAWN BY:		# OF SHEETS:	SHEET #:		SCALE:
PART NAME:		PART #:	MAT:		REQ'D:

NOTES:

Name: _____ Date: _____

Draw the missing section view and complete the title block information.

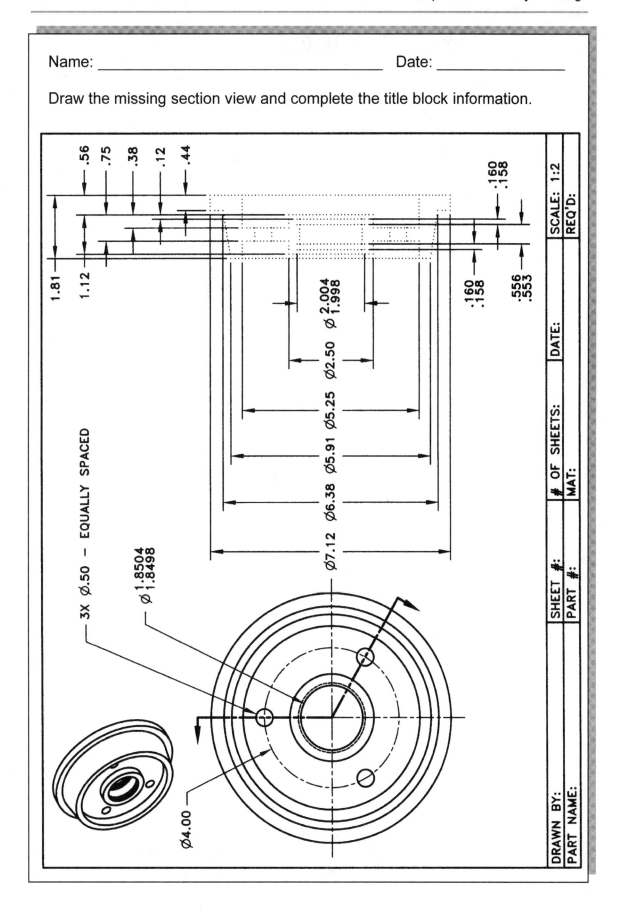

<u>NOTES:</u>

Name: _____   Date: _____

Place the remaining six dimensions and complete the title block information.

.62

.94

$\phi{1.117 \atop 1.116}$

2X 10 – 24 UNC

${2.375 \atop 2.372}$

2X ϕ.88

2.25

1.19

| DRAWN BY: | SHEET #: | # OF SHEETS: | DATE: | SCALE: 1:2 |
| PART NAME: | PART #: | MAT: | | REQ'D: |

NOTES:

Name: _____ Date: _____

Place the remaining nine dimensions and complete the title block information.

NOTES:

Name: _____ Date: _____

Draw in the missing dimensions and complete the title block information.

3X .06 X 45°

Ø .7871 / .7867

Ø .660 / .656

.84

.556 / .553

.104 / .100

R.03

Ø 1.1178 / 1.1175

SCALE: 1:1
REQ'D:
DATE:
OF SHEETS:
MAT:
SHEET #:
PART #:
DRAWN BY:
PART NAME:

NOTES:

Name: _____ Date: _____

Draw and dimension part #5, and fill in the title block.

SCALE: 1:1
REQ'D:

DATE:

OF SHEETS:
MAT:

SHEET #:
PART #:

DRAWN BY:
PART NAME:

NOTES:

Name: _____ Date: _____

Draw and dimension part #6, and fill in the title block.

SCALE: 1:1

REQ'D:

DATE:

OF SHEETS:

MAT:

SHEET #:

PART #:

DRAWN BY:

PART NAME:

NOTES:

Name: _____ Date: _____

Complete the information on the standard (stock) parts and fill in the title block.

7/8 – 9 UNC
HEX HEAD BOLT
LENGTH = 8.25
2 REQ'D
(10)

7/8 – 9 UNC
HEX HEAD NUT
2 REQ'D
(11)

(12)

10 – 24 UNC
ROUND HEAD MACH SCREW
LENGTH = .62
4 REQ'D
(13)

(7)

(8)

(9)

DRAWN BY:
PART NAME: –

SHEET #:
PART #: –

OF SHEETS: –
MAT: –

SCALE: –
REQ'D: –

DATE:

NOTES:

In Class Student Exercise 6-3: Working drawing package 2

Consider the *Drill Jig* shown below. Draw detailed drawings of the individual parts, create a standard parts sheet, and draw an assembly drawing of the *Drill Jig*. The sizes of the individual parts are given on the next page.

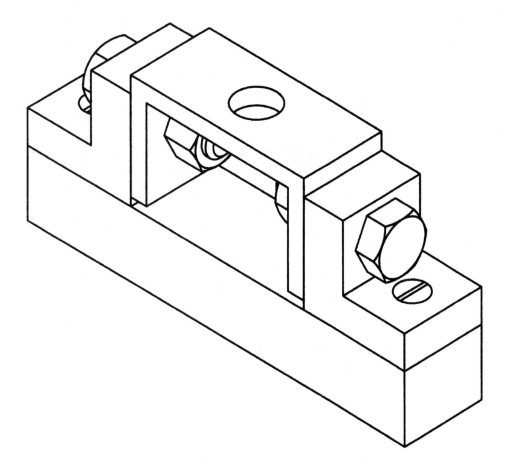

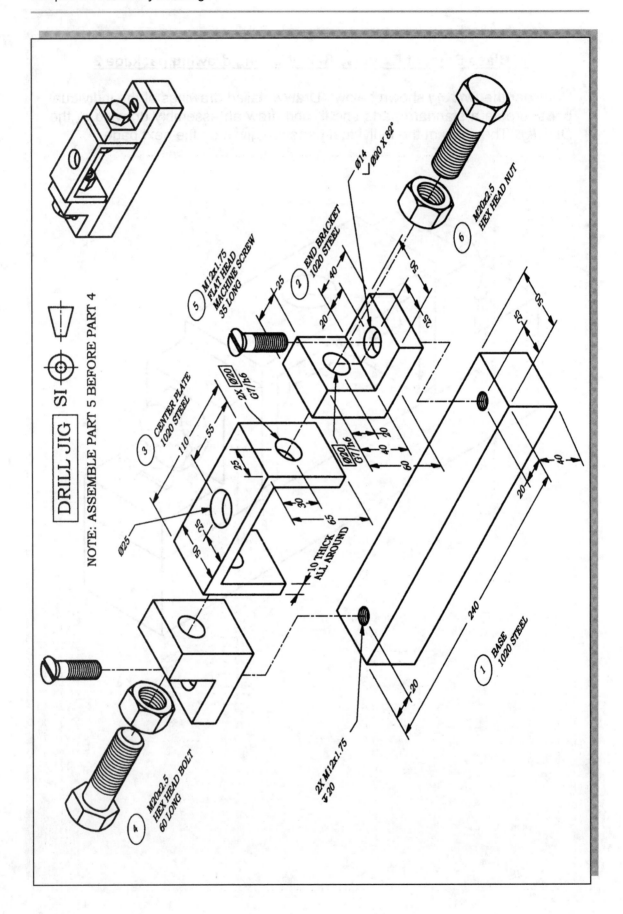

DRILL JIG SI

NOTE: ASSEMBLE PART 5 BEFORE PART 4

1 BASE
1020 STEEL

2 END BRACKET
1020 STEEL

3 CENTER PLATE
1020 STEEL

4 M20x2.5
HEX HEAD BOLT
60 LONG

5 M12x1.75
FLAT HEAD
MACHINE SCREW
35 LONG

6 M20x2.5
HEX HEAD NUT

Name: _____ Date: _____

Complete the section view by adding the appropriate section lines. Then, fill in the part number in the correct balloon, fill in the parts list, and fill in the title block.

NOTES:

Name: _____ Date: _____

Draw and dimension the threaded features of part #1, fill in the title block, and answer the following questions.

- What is the tap drill size for the M12x1.75 thread?
- How much further does the tap drill depth proceed past the thread depth?

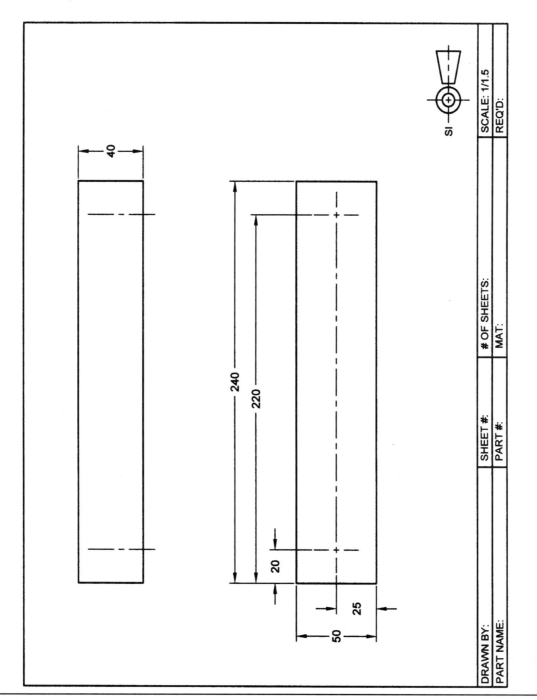

NOTES:

Name: _____ Date: _____

Dimension part #2 and fill in the title block.

SCALE: 1/1.5	REQ'D:
# OF SHEETS:	MAT:
SHEET #:	PART #:
DRAWN BY:	PART NAME:

NOTES:

Name: _____ Date: _____

Draw and dimension part #3, and fill in the title block.

SI

SCALE: 1/1.5

REQ'D:

OF SHEETS:

MAT:

SHEET #:

PART #:

DRAWN BY:

PART NAME:

NOTES:

Name: _____ Date: _____

Create a standard parts sheet and fill in the title block.

SCALE: -

REQ'D: -

OF SHEETS:

MAT: -

SHEET #:

PART #: -

DRAWN BY:

PART NAME: -

<u>NOTES:</u>

ASSEMBLY REVIEW QUESTIONS

Name: _____   Date: _____

Answer the following questions.

Q6-1)   What is the purpose of an assembly drawing?

Q6-2)   What is the difference between a detailed drawing and an assembly drawing?

Q6-3)   What is the first sheet in a working drawing package?

Q6-4)   What is the last sheet in a working drawing package?

Q6-5)   Under what circumstance would we draw section lines in opposing directions?

Q6-6)   Are dimensions usually included in an assembly drawing?

Q6-7)   What criteria determines part number assignment?

Q6-8)   How are parts of an assembly identified?

Q6-9)   What items/headings are usually included in a parts list?

Q6-10)   If we can purchase an item off the shelf, do we need to draw a detail of it?

Q6-11)   What information is contained in a standard parts sheet? Give a one sentence answer, not an itemized list.

ASSEMBLY PROBLEMS

P6-1) Create a working drawing package for the following *Milling Jack*. The working drawing package should contain an assembly drawing, details of all the parts, and a standard parts sheet. Notice that some of the dimensioned isometric drawings are not dimensioned using proper dimensioning technique. When drawing the detailed drawings use proper symbols and dimensioning techniques.

Milling Jack

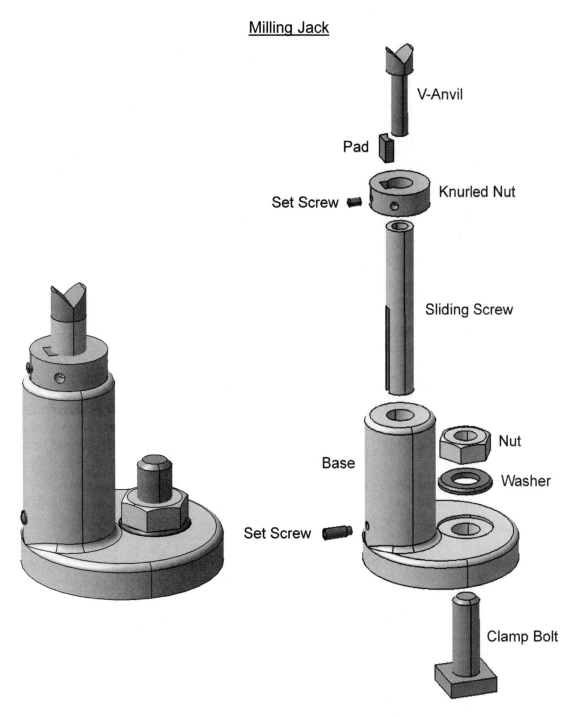

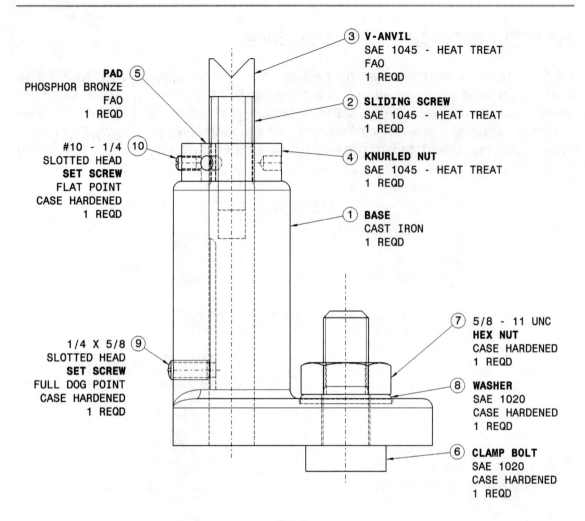

③ **V-ANVIL**
SAE 1045 - HEAT TREAT
FAO
1 REQD

PAD ⑤
PHOSPHOR BRONZE
FAO
1 REQD

② **SLIDING SCREW**
SAE 1045 - HEAT TREAT
1 REQD

#10 - 1/4 ⑩
SLOTTED HEAD
SET SCREW
FLAT POINT
CASE HARDENED
1 REQD

④ **KNURLED NUT**
SAE 1045 - HEAT TREAT
1 REQD

① **BASE**
CAST IRON
1 REQD

1/4 X 5/8 ⑨
SLOTTED HEAD
SET SCREW
FULL DOG POINT
CASE HARDENED
1 REQD

⑦ 5/8 - 11 UNC
HEX NUT
CASE HARDENED
1 REQD

⑧ **WASHER**
SAE 1020
CASE HARDENED
1 REQD

⑥ **CLAMP BOLT**
SAE 1020
CASE HARDENED
1 REQD

Base

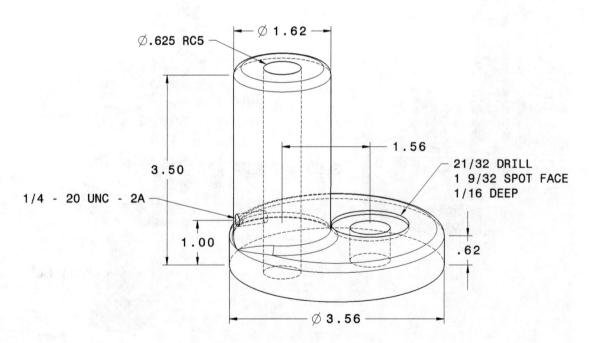

Ø.625 RC5

Ø 1.62

1.56

21/32 DRILL
1 9/32 SPOT FACE
1/16 DEEP

3.50

1/4 - 20 UNC - 2A

1.00

.62

Ø 3.56

Sliding Screw

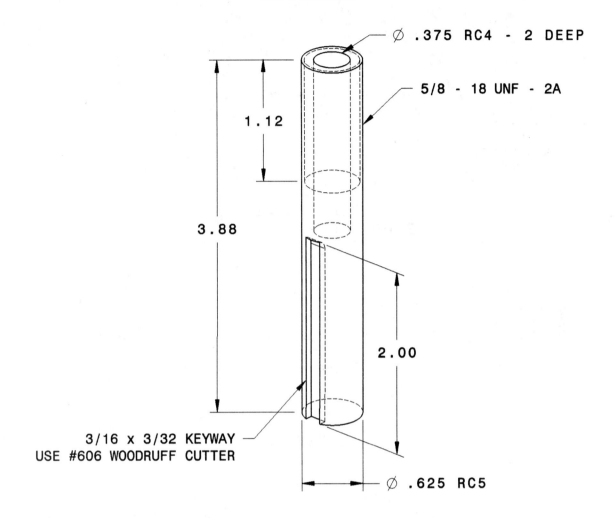

Ø .375 RC4 - 2 DEEP

5/8 - 18 UNF - 2A

1.12

3.88

2.00

3/16 x 3/32 KEYWAY
USE #606 WOODRUFF CUTTER

Ø .625 RC5

V-Anvil

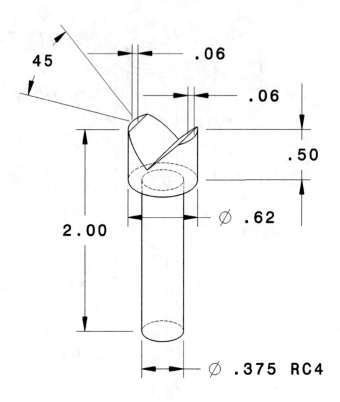

45

.06

.06

.50

2.00

∅ .62

∅ .375 RC4

Knurled Nut

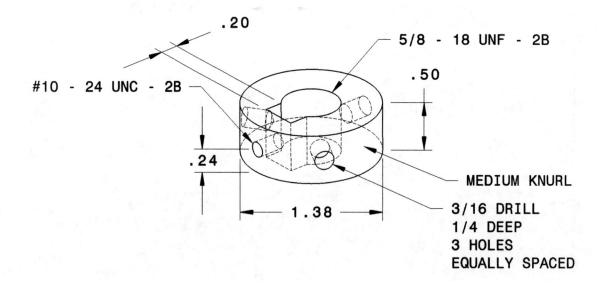

.20

5/8 - 18 UNF - 2B

.50

#10 - 24 UNC - 2B

.24

1.38

MEDIUM KNURL

3/16 DRILL
1/4 DEEP
3 HOLES
EQUALLY SPACED

Pad

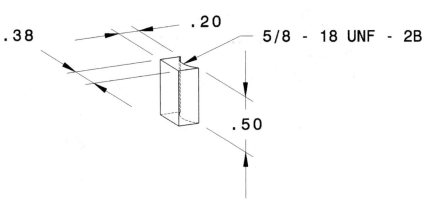

.38 .20 5/8 - 18 UNF - 2B .50

Clamp Bolt

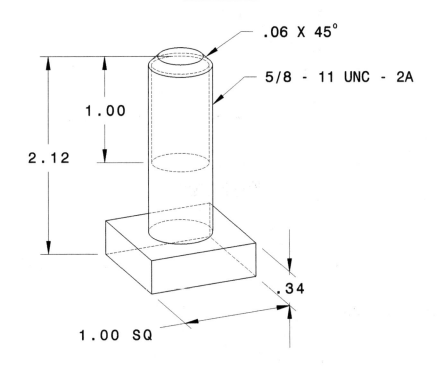

.06 X 45° 5/8 - 11 UNC - 2A 1.00 2.12 .34 1.00 SQ

Washer

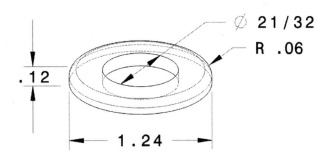

⌀ 21/32 R .06 .12 1.24

NOTES:

APPENDIX A: PRINTING TO SCALE

A.1) PRINTING TO SCALE

Print scale expresses the ratio between the printed size of an object to its actual size. If a drawing is printed full-scale, it implies that a feature dimensioned as 1 inch measures 1 inch with a ruler on the printed drawing. This is referred to as a 1 to 1 scale. Printing full scale, in most cases, is difficult to achieve unless you have access to a plotter. In a classroom setting, most engineering drawings are printed on a standard 8.5" x 11" sheet of paper regardless of the object's size. The scale at which the part is printed out should allow all details and items to be shown clearly and accurately. **Even though a drawing may not be able to be printed full scale, they should always be drawn full scale in the CAD environment.**

Since it is impractical to print all drawings full scale, we employ printing to half scale, quarter-scale and so on. For example, if a drawing is printed half-scale, it means that a feature that is dimensioned 1 inch will measure 0.5 inch on the printed drawing. The scale at which the drawing is printed should be indicated on the drawing next to the text "SCALE" in the title block. On a drawing, half-scale may be denoted in the following ways.

| 1/2 | or | 1:2 | or | 0.5 |

Although it is nice to print to scale, the ASME standard states that no dimension should be measured directly from the printed drawing.

For drawings that are not prepared to any scale, the word "NONE" shall be entered after "SCALE" in the space provided on the drawing.

<u>NOTES:</u>

In Class Student Exercise A-1: Scale

Name: _____ Date: _____

Using the ruler provided, determine the scale that should be indicated on the drawing for the following objects.

Scale =

Scale =

Scale =

NOTES:

APPENDIX B: TITLE BLOCKS

B.1) TITLE BLOCKS

Every engineering drawing should have both a border and a title block. The border defines the drawing area and the title block gives pertinent information about the part or assembly being drawn. There are several different types of title blocks, but they all contain similar information. The information that is included depends on the drawing type, field of engineering, and viewing audience.

B.2) TITLE BLOCK CONTENT

The information contained in a title block may include, but is not limited to, the following:

1. Name of drafter
2. Checked by
3. School or Company
4. Drawing title
5. Part name
6. Part number
7. Material of part
8. Number of required parts
9. Sheet number
10. Number of sheets
11. Scale of drawing
12. Date
13. Last revision

B.2.1) Date

The drawing date is given numerically in order of *year-month-day*. For example, the date, May 31, 2005, would be indicated as 2005-05-31 or 2005/05/31. The date is placed in the title block next to the word "DATE".

B.3) SHEET LAYOUT

In a class room setting most drawings are printed out on an 8.5" x 11" sheet of paper. Figure B-1 shows a typical drawing sheet layout. Figures B-2 through B-4 show three different title block forms.

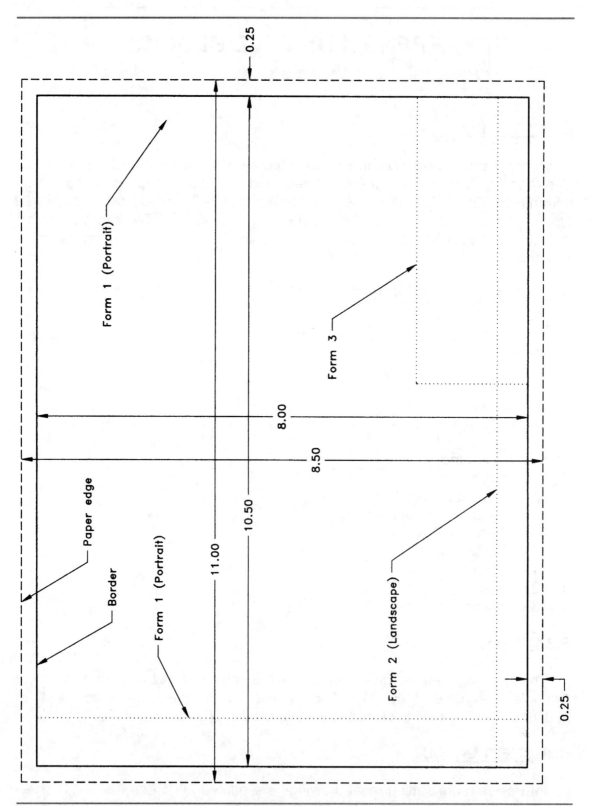

Figure B-1: Sheet layout for a 11 x 8.5 sheet of paper.

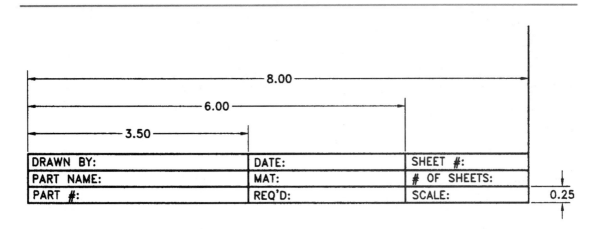

Figure B-2: Form 1 (portrait title block)

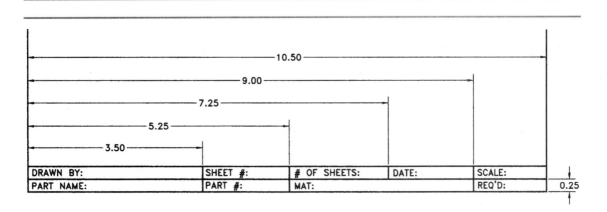

Figure B-3: Form 2 (Landscape title block)

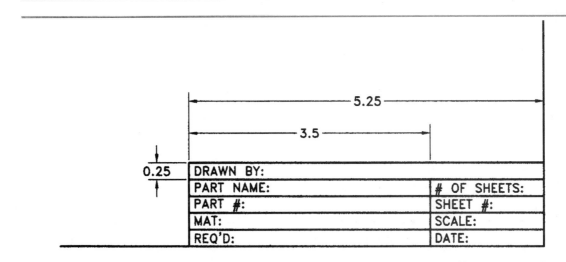

Figure B-4: Form 3 (title block)

NOTES:

APPENDIX C: SYMBOL FORM AND PROPORTION

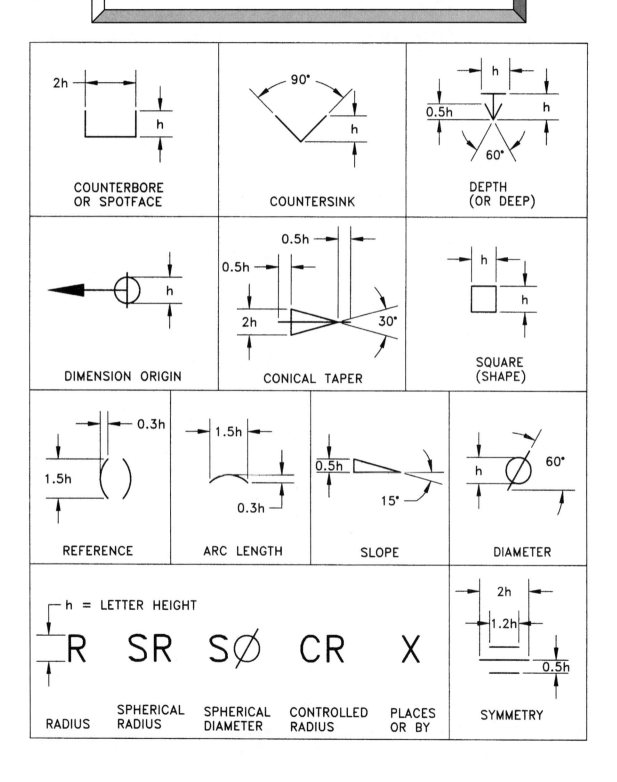

NOTES:

APPENDIX D: LIMITS AND FITS

D.1) LIMITS AND FITS (INCH)

D.1.1) Running and Sliding Fits

Basic hole system. Limits are in thousandths of an inch.
Limits for hole and shaft are applied algebraically to the basic size to obtain the limits of size for the parts.

Nominal Size Range Inches		Class RC1 Standard Limits		Class RC2 Standard Limits		Class RC3 Standard Limits		Class RC4 Standard Limits	
Over	To	Hole	Shaft	Hole	Shaft	Hole	Shaft	Hole	Shaft
0	− 0.12	+0.2 0	-0.1 -0.25	+0.25 0	-0.1 -0.3	+0.4 0	-0.3 -0.55	+0.6 0	-0.3 -0.7
0.12	− 0.24	+0.2 0	-0.15 -0.3	+0.3 0	-0.15 -0.35	+0.5 0	-0.4 -0.7	+0.7 0	-0.4 -0.9
0.24	− 0.40	+0.25 0	-0.2 -0.35	+0.4 0	-0.2 -0.45	+0.6 0	-0.5 -0.9	+0.9 0	-0.5 -1.1
0.40	− 0.71	+0.3 0	-0.25 -0.45	+0.4 0	-0.25 -0.55	+0.7 0	-0.6 -1.0	+1.0 0	-0.6 -1.3
0.71	− 1.19	+0.4 0	-0.3 -0.55	+0.5 0	-0.3 -0.7	+0.8 0	-0.8 -1.3	+1.2 0	-0.8 -1.6
1.19	− 1.97	+0.4 0	-0.4 -0.7	+0.6 0	-0.4 -0.8	+1.0 0	-1.0 -1.6	+1.6 0	-1.0 -2.0
1.97	− 3.15	+0.5 0	-0.4 -0.7	+0.7 0	-0.4 -0.9	+1.2 0	-1.2 -1.9	+1.8 0	-1.2 -2.4
3.15	− 4.73	+0.6 0	-0.5 -0.9	+0.9 0	-0.5 -1.1	+1.4 0	-1.4 -2.3	+2.2 0	-1.4 -2.8
4.73	− 7.09	+0.7 0	-0.6 -1.1	+1.0 0	-0.6 -1.3	+1.6 0	-1.6 -2.6	+2.5 0	-1.6 -3.2
7.09	− 9.85	+0.8 0	-0.6 -1.2	+1.2 0	-0.6 -1.4	+1.8 0	-2.0 -3.2	+2.8 0	-2.0 -3.8
9.85	− 12.41	+0.9 0	-0.8 -1.4	+1.2 0	-0.7 -1.6	+2.0 0	-2.5 -3.7	+3.0 0	-2.2 -4.2
12.41	− 15.75	+1.0 0	-1.0 -1.7	+1.4 0	-0.7 -1.7	+2.2 0	-3.0 -4.4	+3.5 0	-2.5 -4.7
15.75	− 19.69	+1.0 0	-1.2 -2.0	+1.6 0	-0.8 -1.8	+2.5 0	-4.0 -5.6	+4.0 0	-2.8 -5.3

Nominal Size Range Inches		Class RC5		Class RC6		Class RC7		Class RC8		Class RC9	
		Standard Limits		Standard Limits		Standard Limits		Standard Limits		Standard Limits	
Over	To	Hole	Shaft	Hole	Shaft	Hole	Shaft	Hole	Shaft	Hole	Shaft
0	− 0.12	+0.6	-0.6	+1.0	-0.6	+1.0	-1.0	+1.6	-2.5	+2.5	-4.0
		0	-1.0	0	-1.2	0	-1.6	0	-3.5	0	-5.6
0.12	− 0.24	+0.7	-0.8	+1.2	-0.8	+1.2	-1.2	+1.8	-2.8	+3.0	-4.5
		0	-1.3	0	-1.5	0	-1.9	0	-4.0	0	-6.0
0.24	− 0.40	+0.9	-1.0	+1.4	-1.0	+1.4	-1.6	+2.2	-3.0	+3.5	-5.0
		0	-1.6	0	-1.9	0	-2.5	0	-4.4	0	-7.2
0.40	− 0.71	+1.0	-1.2	+1.6	-1.2	+1.6	-2.0	+2.8	-3.5	+4.0	-6.0
		0	-1.9	0	-2.2	0	-3.0	0	-5.1	0	-8.8
0.71	− 1.19	+1.2	-1.6	+2.0	-1.6	+2.0	-2.5	+3.5	-4.5	+5.0	-7.0
		0	-2.4	0	-2.8	0	-3.7	0	-6.5	0	-10.5
1.19	− 1.97	+1.6	-2.0	+2.5	-2.0	+2.5	-3.0	+4.0	-5.0	+6.0	-8.0
		0	-3.0	0	-3.6	0	-4.6	0	-7.5	0	12.0
1.97	− 3.15	+1.8	-2.5	+3.0	-2.5	+3.0	-4.0	+4.5	-6.0	+7.0	-9.0
		0	-3.7	0	-4.3	0	-5.8	0	-9.0	0	-13.5
3.15	− 4.73	+2.2	-3.0	+3.5	-3.0	+3.5	-5.0	+5.0	-7.0	+9.0	-10.0
		0	-4.4	0	-5.2	0	-7.2	0	-10.5	0	-15.0
4.73	− 7.09	+2.5	-3.5	+4.0	-3.5	+4.0	-6.0	+6.0	-8.0	+10.0	-12.0
		0	-5.1	0	-6.0	0	-8.5	0	-12.0	0	-18.0
7.09	− 9.85	+2.8	-4.0	+4.5	-4.0	+4.5	-7.0	+7.0	-10.0	+12.0	-15.0
		0	-5.8	0	-6.8	0	-9.8	0	-14.5	0	-22.0
9.85	− 12.41	+3.0	-5.0	+5.0	-5.0	+5.0	-8.0	+8.0	-12.0	+12.0	-18.0
		0	-7.0	0	-8.0	0	-11.0	0	-17.0	0	-26.0
12.41	− 15.75	+3.5	-6.0	+6.0	-6.0	+6.0	-10.0	+9.0	-14.0	+14.0	-22.0
		0	-8.2	0	-9.5	0	-13.5	0	-20.0	0	-31.0
15.75	− 19.69	+4.0	-8.0	+6.0	-8.0	+6.0	-12.0	+10.0	-16.0	+16.0	-25.0
		0	-10.5	0	-12.0	0	-16.0	0	-22.0	0	-35.0

USAS/ASME B4.1 − 1967 (R2004) Standard. For larger diameters, see the standard. ASME/ANSI B18.3.5M − 1986 (R2002) Standard. Reprinted from the standard listed by permission of the American Society of Mechanical Engineers. All rights reserved.

D.1.2) Locational Clearance Fits

Basic hole system. Limits are in thousandths of an inch.
Limits for hole and shaft are applied algebraically to the basic size to obtain the limits of size for the parts.

Nominal Size Range Inches		Class LC1 Standard Limits		Class LC2 Standard Limits		Class LC3 Standard Limits		Class LC4 Standard Limits	
Over	To	Hole	Shaft	Hole	Shaft	Hole	Shaft	Hole	Shaft
0	− 0.12	+0.25 / 0	0 / -0.2	+0.4 / 0	0 / -0.25	+0.6 / 0	0 / -0.4	+1.6 / 0	0 / -1.0
0.12	− 0.24	+0.3 / 0	0 / -0.2	+0.5 / 0	0 / -0.3	+0.7 / 0	0 / -0.5	+1.8 / 0	0 / -1.2
0.24	− 0.40	+0.4 / 0	0 / -0.25	+0.6 / 0	0 / -0.4	+0.9 / 0	0 / -0.6	+2.2 / 0	0 / -1.4
0.40	− 0.71	+0.4 / 0	0 / -0.3	+0.7 / 0	0 / -0.4	+1.0 / 0	0 / -0.7	+2.8 / 0	0 / -1.6
0.71	− 1.19	+0.5 / 0	0 / -0.4	+0.8 / 0	0 / -0.5	+1.2 / 0	0 / -0.8	+3.5 / 0	0 / -2.0
1.19	− 1.97	+0.6 / 0	0 / -0.4	+1.0 / 0	0 / -0.6	+1.6 / 0	0 / -1.0	+4.0 / 0	0 / -2.5
1.97	− 3.15	+0.7 / 0	0 / -0.5	+1.2 / 0	0 / -0.7	+1.8 / 0	0 / -1.2	+4.5 / 0	0 / -3.0
3.15	− 4.73	+0.9 / 0	0 / -0.6	+1.4 / 0	0 / -0.9	+2.2 / 0	0 / -1.4	+5.0 / 0	0 / -3.5
4.73	− 7.09	+1.0 / 0	0 / -0.7	+1.6 / 0	0 / -1.0	+2.5 / 0	0 / -1.6	+6.0 / 0	0 / -4.0
7.09	− 9.85	+1.2 / 0	0 / -0.8	+1.8 / 0	0 / -1.2	+2.8 / 0	0 / -1.8	+7.0 / 0	0 / -4.5
9.85	− 12.41	+1.2 / 0	0 / -0.9	+2.0 / 0	0 / -1.2	+3.0 / 0	0 / -2.0	+8.0 / 0	0 / -5.0
12.41	− 15.75	+1.4 / 0	0 / -1.0	+2.2 / 0	0 / -1.4	+3.5 / 0	0 / -2.2	+9.0 / 0	0 / -6.0
15.75	− 19.69	+1.6 / 0	0 / -1.0	+2.5 / 0	0 / -1.6	+4.0 / 0	0 / -2.5	+10.0 / 0	0 / -6.0

Nominal Size Range Inches		Class LC5 Standard Limits		Class LC6 Standard Limits		Class LC7 Standard Limits		Class LC8 Standard Limits	
Over	To	Hole	Shaft	Hole	Shaft	Hole	Shaft	Hole	Shaft
0	− 0.12	+0.4 / 0	-0.1 / -0.35	+1.0 / 0	-0.3 / -0.9	+1.6 / 0	-0.6 / -1.6	+1.6 / 0	-1.0 / -2.0
0.12	− 0.24	+0.5 / 0	-0.15 / -0.45	+1.2 / 0	-0.4 / -1.1	+1.8 / 0	-0.8 / -2.0	+1.8 / 0	-1.2 / -2.4
0.24	− 0.40	+0.6 / 0	-0.2 / -0.6	+1.4 / 0	-0.5 / -1.4	+2.2 / 0	-1.0 / -2.4	+2.2 / 0	-1.6 / -3.0
0.40	− 0.71	+0.7 / 0	-0.25 / -0.65	+1.6 / 0	-0.6 / -1.6	+2.8 / 0	-1.2 / -2.8	+2.8 / 0	-2.0 / -3.6
0.71	− 1.19	+0.8 / 0	-0.3 / -0.8	+2.0 / 0	-0.8 / -2.0	+3.5 / 0	-1.6 / -3.6	+3.5 / 0	-2.5 / -4.5
1.19	− 1.97	+1.0 / 0	-0.4 / -1.0	+2.5 / 0	-1.0 / -2.6	+4.0 / 0	-2.0 / -4.5	+4.0 / 0	-3.0 / -5.5
1.97	− 3.15	+1.2 / 0	-0.4 / -1.1	+3.0 / 0	-1.2 / -3.0	+4.5 / 0	-2.5 / -5.5	+4.5 / 0	-4.0 / -7.0
3.15	− 4.73	+1.4 / 0	-0.5 / -1.4	+3.5 / 0	-1.4 / -3.6	+5.0 / 0	-3.0 / -6.5	+5.0 / 0	-5.0 / -8.5
4.73	− 7.09	+1.6 / 0	-0.6 / -1.6	+4.0 / 0	-1.6 / -4.1	+6.0 / 0	-3.5 / -7.5	+6.0 / 0	-6.0 / -10.0
7.09	− 9.85	+1.8 / 0	-0.6 / -1.8	+4.5 / 0	-2.0 / -4.8	+7.0 / 0	-4.0 / -8.5	+7.0 / 0	-7.0 / -11.5
9.85	− 12.41	+2.0 / 0	-0.7 / -1.9	+5.0 / 0	-2.2 / -5.2	+8.0 / 0	-4.5 / -9.5	+8.0 / 0	-7.0 / -12.0
12.41	− 15.75	+2.2 / 0	-0.7 / -2.1	+6.0 / 0	-2.5 / -6.0	+9.0 / 0	-5.0 / -11.0	+9.0 / 0	-8.0 / -14.0
15.75	− 19.69	+2.5 / 0	-0.8 / -2.4	+6.0 / 0	-2.8 / -6.8	+10.0 / 0	-5.0 / -11.0	+10.0 / 0	-9.0 / -15.0

Nominal Size Range Inches		Class LC9 Standard Limits		Class LC10 Standard Limits		Class LC11 Standard Limits	
Over	To	Hole	Shaft	Hole	Shaft	Hole	Shaft
0	− 0.12	+2.5	-2.5	+4.0	-4.0	+6.0	-5.0
		0	-4.1	0	-8.0	0	-11.0
0.12	− 0.24	+3.0	-2.8	+5.0	-4.5	+7.0	-6.0
		0	-4.6	0	-9.5	0	-13.0
0.24	− 0.40	+3.5	-3.0	+6.0	-5.0	+9.0	-7.0
		0	-5.2	0	-11.0	0	-16.0
0.40	− 0.71	+4.0	-3.5	+7.0	-6.0	+10.0	-8.0
		0	-6.3	0	-13.0	0	-18.0
0.71	− 1.19	+5.0	-4.5	+8.0	-7.0	+12.0	-10.0
		0	-8.0	0	-15.0	0	-22.0
1.19	− 1.97	+6.0	-5.0	+10.0	-8.0	+16.0	-12.0
		0	-9.0	0	-18.0	0	-28.0
1.97	− 3.15	+7.0	-6.0	+12.0	-10.0	+18.0	-14.0
		0	-10.5	0	-22.0	0	-32.0
3.15	− 4.73	+9.0	-7.0	+14.0	-11.0	+22.0	-16.0
		0	-12.0	0	-25.0	0	-38.0
4.73	− 7.09	+10.0	-8.0	+16.0	-12.0	+25.0	-18.0
		0	-14.0	0	-28.0	0	-43.0
7.09	− 9.85	+12.0	-10.0	+18.0	-16.0	+28.0	-22.0
		0	-17.0	0	-34.0	0	-50.0
9.85	− 12.41	+12.0	-12.0	+20.0	-20.0	+30.0	-28.0
		0	-20.0	0	-40.0	0	-58.0
12.41	− 15.75	+14.0	-14.0	+22.0	-22.0	+35.0	-30.0
		0	-23.0	0	-44.0	0	-65.0
15.75	− 19.69	+16.0	-16.0	+25.0	-25.0	+40.0	-35.0
		0	-26.0	0	-50.0	0	-75.0

USAS/ASME B4.1 – 1967 (R2004) Standard. For larger diameters, see the standard. ASME/ANSI B18.3.5M – 1986 (R2002) Standard. Reprinted from the standard listed by permission of the American Society of Mechanical Engineers. All rights reserved.

D.1.3) __Locational Transition Fits__

Basic hole system. Limits are in thousandths of an inch.
Limits for hole and shaft are applied algebraically to the basic size to obtain the limits of size for the parts.

Nominal Size Range Inches		Class LT1 Standard Limits		Class LT2 Standard Limits		Class LT3 Standard Limits	
Over	To	Hole	Shaft	Hole	Shaft	Hole	Shaft
0	− 0.12	+0.4 / 0	+0.10 / -0.10	+0.6 / 0	+0.2 / -0.2		
0.12	− 0.24	+0.5 / 0	+0.15 / -0.15	+0.7 / 0	+0.25 / -0.25		
0.24	− 0.40	+0.6 / 0	+0.2 / -0.2	+0.9 / 0	+0.3 / -0.3	+0.6 / 0	+0.5 / +0.1
0.40	− 0.71	+0.7 / 0	+0.2 / -0.2	+1.0 / 0	+0.35 / -0.35	+0.7 / 0	+0.5 / +0.1
0.71	− 1.19	+0.8 / 0	+0.25 / -0.25	+1.2 / 0	+0.4 / -0.4	+0.8 / 0	+0.6 / +0.1
1.19	− 1.97	+1.0 / 0	+0.3 / -0.3	+1.6 / 0	+0.5 / -0.5	+1.0 / 0	+0.7 / +0.1
1.97	− 3.15	+1.2 / 0	+0.3 / -0.3	+1.8 / 0	+0.6 / -0.6	+1.2 / 0	+0.8 / +0.1
3.15	− 4.73	+1.4 / 0	+0.4 / -0.4	+2.2 / 0	+0.7 / -0.7	+1.4 / 0	+1.0 / +0.1
4.73	− 7.09	+1.6 / 0	+0.5 / -0.5	+2.5 / 0	+0.8 / -0.8	+1.6 / 0	+1.1 / +0.1
7.09	− 9.85	+1.8 / 0	+0.6 / -0.6	+2.8 / 0	+0.9 / -0.9	+1.8 / 0	+1.4 / +0.2
9.85	− 12.41	+2.0 / 0	+0.6 / -0.6	+3.0 / 0	+1.0 / -1.0	+2.0 / 0	+1.4 / +0.2
12.41	− 15.75	+2.2 / 0	+0.7 / -0.7	+3.5 / 0	+1.0 / -1.0	+2.2 / 0	+1.6 / +0.2

Nominal Size Range Inches		Class LT4 Standard Limits		Class LT5 Standard Limits		Class LT6 Standard Limits	
Over	To	Hole	Shaft	Hole	Shaft	Hole	Shaft
0	− 0.12			+0.4 / 0	+0.5 / +0.25	+0.4 / 0	-0.65 / +0.25
0.12	− 0.24			+0.5 / 0	+0.6 / +0.3	+0.5 / 0	+0.8 / +0.3
0.24	− 0.40	+0.9 / 0	+0.7 / +0.1	+0.6 / 0	+0.8 / +0.4	+0.6 / 0	+1.0 / +0.4
0.40	− 0.71	+1.0 / 0	+0.8 / +0.1	+0.7 / 0	+0.9 / +0.5	+0.7 / 0	+1.2 / +0.5
0.71	− 1.19	+1.2 / 0	+0.9 / +0.1	+0.8 / 0	+1.1 / +0.6	+0.8 / 0	+1.4 / +0.6
1.19	− 1.97	+1.6 / 0	+1.1 / +0.1	+1.0 / 0	+1.3 / +0.7	+1.0 / 0	+1.7 / +0.7
1.97	− 3.15	+1.8 / 0	+1.3 / +0.1	+1.2 / 0	+1.5 / +0.8	+1.2 / 0	+2.0 / +0.8
3.15	− 4.73	+2.2 / 0	+1.5 / +0.1	+1.4 / 0	+1.9 / +1.0	+1.4 / 0	+2.4 / +1.0
4.73	− 7.09	+2.5 / 0	+1.7 / +0.1	+1.6 / 0	+2.2 / +1.2	+1.6 / 0	+2.8 / +1.2
7.09	− 9.85	+2.8 / 0	+2.0 / +0.2	+1.8 / 0	+2.6 / +1.4	+1.8 / 0	+3.2 / +1.4
9.85	− 12.41	+3.0 / 0	+2.2 / +0.2	+2.0 / 0	+2.6 / +1.4	+2.0 / 0	+3.4 / +1.4
12.41	− 15.75	+3.5 / 0	+2.4 / +0.2	+2.2 / 0	+3.0 / +1.6	+2.2 / 0	+3.8 / +1.6

USAS/ASME B4.1 − 1967 (R2004) Standard. For larger diameters, see the standard. ASME/ANSI B18.3.5M − 1986 (R2002) Standard. Reprinted from the standard listed by permission of the American Society of Mechanical Engineers. All rights reserved.

D.1.4) Locational Interference Fits

Basic hole system. Limits are in thousandths of an inch.
Limits for hole and shaft are applied algebraically to the basic size to obtain the limits of size for the parts.

Nominal Size Range Inches		Class LN1 Standard Limits		Class LN2 Standard Limits		Class LN3 Standard Limits	
Over	To	Hole	Shaft	Hole	Shaft	Hole	Shaft
0	− 0.12	+0.25	+0.45	+0.4	+0.65	+0.4	+0.75
		0	+0.25	0	+0.4	0	+0.5
0.12	− 0.24	+0.3	+0.5	+0.5	+0.8	+0.5	+0.9
		0	+0.3	0	+0.5	0	+0.6
0.24	− 0.40	+0.4	+0.65	+0.6	+1.0	+0.6	+1.2
		0	+0.4	0	+0.6	0	+0.8
0.40	− 0.71	+0.4	+0.8	+0.7	+1.1	+0.7	+1.4
		0	+0.4	0	+0.7	0	+1.0
0.71	− 1.19	+0.5	+1.0	+0.8	+1.3	+0.8	+1.7
		0	+0.5	0	+0.8	0	+1.2
1.19	− 1.97	+0.6	+1.1	+1.0	+1.6	+1.0	+2.0
		0	+0.6	0	+1.0	0	+1.4
1.97	− 3.15	+0.7	+1.3	+1.2	+2.1	+1.2	+2.3
		0	+0.8	0	+1.4	0	+1.6
3.15	− 4.73	+0.9	+1.6	+1.4	+2.5	+1.4	+2.9
		0	+1.0	0	+1.6	0	+2.0
4.73	− 7.09	+1.0	+1.9	+1.6	+2.8	+1.6	+3.5
		0	+1.2	0	+1.8	0	+2.5
7.09	− 9.85	+1.2	+2.2	+1.8	+3.2	+1.8	+4.2
		0	+1.4	0	+2.0	0	+3.0
9.85	− 12.41	+1.2	+2.3	+2.0	+3.4	+2.0	+4.7
		0	+1.4	0	+2.2	0	+3.5
12.41	− 15.75	+1.4	+2.6	+2.2	+3.9	+2.2	+5.9
		0	+1.6	0	+2.5	0	+4.5
15.75	− 19.69	+1.6	+2.8	+2.5	+4.4	+2.5	+6.6
		0	+1.8	0	+2.8	0	+5.0

USAS/ASME B4.1 − 1967 (R2004) Standard. For larger diameters, see the standard. ASME/ANSI B18.3.5M − 1986 (R2002) Standard. Reprinted from the standard listed by permission of the American Society of Mechanical Engineers. All rights reserved.

D.1.5) Force and Shrink Fits

Basic hole system. Limits are in thousandths of an inch.
Limits for hole and shaft are applied algebraically to the basic size to obtain the limits of size for the parts.

Nominal Size Range Inches		Class FN1		Class FN2		Class FN3		Class FN4		Class FN5	
		Standard Limits		Standard Limits		Standard Limits		Standard Limits		Standard Limits	
Over	To	Hole	Shaft	Hole	Shaft	Hole	Shaft	Hole	Shaft	Hole	Shaft
0	− 0.12	+0.25 / 0	+0.5 / +0.3	+0.4 / 0	+0.85 / +0.6			+0.4 / 0	+0.95 / +0.7	+0.6 / 0	+1.3 / +0.9
0.12	− 0.24	+0.3 / 0	+0.6 / +0.4	+0.5 / 0	+1.0 / +0.7			+0.5 / 0	+1.2 / +0.9	+0.7 / 0	+1.7 / +1.2
0.24	− 0.40	+0.4 / 0	+0.75 / +0.5	+0.6 / 0	+1.4 / +1.0			+0.6 / 0	+1.6 / +1.2	+0.9 / 0	+2.0 / +1.4
0.40	− 0.56	+0.4 / 0	+0.8 / +0.5	+0.7 / 0	+1.6 / +1.2			+0.7 / 0	+1.8 / +1.4	+1.0 / 0	+2.3 / +1.6
0.56	− 0.71	+0.4 / 0	+0.9 / +0.6	+0.7 / 0	+1.6 / +1.2			+0.7 / 0	+1.8 / +1.4	+1.0 / 0	+2.5 / +1.8
0.71	− 0.95	+0.5 / 0	+1.1 / +0.7	+0.8 / 0	+1.9 / +1.4			+0.8 / 0	+2.1 / +1.6	+1.2 / 0	+3.0 / +2.2
0.95	− 1.19	+0.5 / 0	+1.2 / +0.8	+0.8 / 0	+1.9 / +1.4	+0.8 / 0	+2.1 / +1.6	+0.8 / 0	+2.3 / +1.8	+1.2 / 0	+3.3 / +2.5
1.19	− 1.58	+0.6 / 0	+1.3 / +0.9	+1.0 / 0	+2.4 / +1.8	+1.0 / 0	+2.6 / +2.0	+1.0 / 0	+3.1 / +2.5	+1.6 / 0	+4.0 / +3.0
1.58	− 1.97	+0.6 / 0	+1.4 / +1.0	+1.0 / 0	+2.4 / +1.8	+1.0 / 0	+2.8 / +2.2	+1.0 / 0	+3.4 / +2.8	+1.6 / 0	+5.0 / +4.0
1.97	− 2.56	+0.7 / 0	+1.8 / +1.3	+1.2 / 0	+2.7 / +2.0	+1.2 / 0	+3.2 / +2.5	+1.2 / 0	+4.2 / +3.5	+1.8 / 0	+6.2 / +5.0
2.56	− 3.15	+0.7 / 0	+1.9 / +1.4	+1.2 / 0	+2.9 / +2.2	+1.2 / 0	+3.7 / +3.0	+1.2 / 0	+4.7 / +4.0	+1.8 / 0	+7.2 / +6.0
3.15	− 3.94	+0.9 / 0	+2.4 / +1.8	+1.4 / 0	+3.7 / +2.8	+1.4 / 0	+4.4 / +3.5	+1.4 / 0	+5.9 / +5.0	+2.2 / 0	+8.4 / +7.0
3.94	− 4.73	+0.9 / 0	+2.6 / +2.0	+1.4 / 0	+3.9 / +3.0	+1.4 / 0	+4.9 / +4.0	+1.4 / 0	+6.9 / +6.0	+2.2 / 0	+9.4 / +8.0
4.73	−5.52	+1.0 / 0	+2.9 / +2.2	+1.6 / 0	+4.5 / +3.5	+1.6 / 0	+6.0 / +5.0	+1.6 / 0	+8.0 / +7.0	+2.5 / 0	+11.6 / +10.0
5.52	−6.30	+1.0 / 0	+3.2 / +2.5	+1.6 / 0	+5.0 / +4.0	+1.6 / 0	+6.0 / +5.0	+1.6 / 0	+8.0 / +7.0	+2.5 / 0	+13.6 / +12.0
6.30	−7.09	+1.0 / 0	+3.5 / +2.8	+1.6 / 0	+5.5 / +4.5	+1.6 / 0	+7.0 / +6.0	+1.6 / 0	+9.0 / +8.0	+2.5 / 0	+13.6 / +12.0
7.09	−7.88	+1.2 / 0	+3.8 / +3.0	+1.8 / 0	+6.2 / +5.0	+1.8 / 0	+8.2 / +7.0	+1.8 / 0	+10.2 / +9.0	+2.8 / 0	+15.8 / +14.0
7.88	−8.86	+1.2 / 0	+4.3 / +3.5	+1.8 / 0	+6.2 / +5.0	+1.8 / 0	+8.2 / +7.0	+1.8 / 0	+11.2 / +10.0	+2.8 / 0	+17.8 / +16.0
8.86	−9.86	+1.2 / 0	+4.3 / +3.5	+1.8 / 0	+7.2 / +6.0	+1.8 / 0	+9.2 / +8.0	+1.8 / 0	+13.2 / +12.0	+2.8 / 0	+17.8 / +16.0
9.85	−11.03	+1.2 / 0	+4.9 / +4.0	+2.0 / 0	+7.2 / +6.0	+2.0 / 0	+10.2 / +9.0	+2.0 / 0	+13.2 / +12.0	+3.0 / 0	+20.0 / +18.0

USAS/ASME B4.1 – 1967 (R2004) Standard. For larger diameters, see the standard. ASME/ANSI B18.3.5M – 1986 (R2002) Standard. Reprinted from the standard listed by permission of the American Society of Mechanical Engineers. All rights reserved.

D.2) METRIC LIMITS AND FITS

D.2.1) Hole Basis Clearance Fits

Preferred Hole Basis Clearance Fits. Dimensions in mm.

Basic Size	Loose Running		Free Running		Close Running		Sliding		Locational Clearance	
	Hole H11	Shaft c11	Hole H9	Shaft d9	Hole H8	Shaft f7	Hole H7	Shaft g6	Hole H7	Shaft h6
1 max	1.060	0.940	1.025	0.980	1.014	0.994	1.010	0.998	1.010	1.000
min	1.000	0.880	1.000	0.955	1.000	0.984	1.000	0.992	1.000	0.994
1.2 max	1.260	1.140	1.225	1.180	1.214	1.194	1.210	1.198	1.210	1.200
min	1.200	1.080	1.200	1.155	1.200	1.184	1.200	1.192	1.200	1.194
1.6 max	1.660	1.540	1.625	1.580	1.614	1.594	1.610	1.598	1.610	1.600
min	1.600	1.480	1.600	1.555	1.600	1.584	1.600	1.592	1.600	1.594
2 max	2.060	1.940	2.025	1.980	2.014	1.994	2.010	1.998	2.010	2.000
min	2.000	1.880	2.000	1.955	2.000	1.984	2.000	1.992	2.000	1.994
2.5 max	2.560	2.440	2.525	2.480	2.514	2.494	2.510	2.498	2.510	2.500
min	2.500	2.380	2.500	2.455	2.500	2.484	2.500	2.492	2.500	2.494
3 max	3.060	2.940	3.025	2.980	3.014	2.994	3.010	2.998	3.010	3.000
min	3.000	2.880	3.000	2.955	3.000	2.984	3.000	2.992	3.000	2.994
4 max	4.075	3.930	4.030	3.970	4.018	3.990	4.012	3.996	4.012	4.000
min	4.000	3.855	4.000	3.940	4.000	3.978	4.000	3.988	4.000	3.992
5 max	5.075	4.930	5.030	4.970	5.018	4.990	5.012	4.996	5.012	5.000
min	5.000	4.855	5.000	4.940	5.000	4.978	5.000	4.988	5.000	4.992
6 max	6.075	5.930	6.030	5.970	6.018	5.990	6.012	5.996	6.012	6.000
min	6.000	5.855	6.000	5.940	6.000	5.978	6.000	5.988	6.000	5.992
8 max	8.090	7.920	8.036	7.960	8.022	7.987	8.015	7.995	8.015	8.000
min	8.000	7.830	8.000	7.924	8.000	7.972	8.000	7.986	8.000	7.991
10 max	10.090	9.920	10.036	9.960	10.022	9.987	10.015	9.995	10.015	10.000
min	10.000	9.830	10.000	9.924	10.000	9.972	10.000	9.986	10.000	9.991
12 max	12.110	11.905	12.043	11.950	12.027	11.984	12.018	11.994	12.018	12.000
min	12.000	11.795	12.000	11.907	12.000	11.966	12.000	11.983	12.000	11.989
16 max	16.110	15.905	16.043	15.950	16.027	15.984	16.018	15.994	16.018	16.000
min	16.000	15.795	16.000	15.907	16.000	15.966	16.000	15.983	16.000	15.989
20 max	20.130	19.890	20.052	19.935	20.033	19.980	20.021	19.993	20.021	20.000
min	20.000	19.760	20.000	19.883	20.000	19.959	20.000	19.980	20.000	19.987
25 max	25.130	24.890	25.052	24.935	25.033	24.980	25.021	24.993	25.021	25.000
min	25.000	24.760	25.000	24.883	25.000	24.959	25.000	24.980	25.000	24.987
30 max	30.130	29.890	30.052	29.935	30.033	29.980	30.021	29.993	30.021	30.000
min	30.000	29.760	30.000	29.883	30.000	29.959	30.000	29.980	30.000	29.987

ANSI B4.2 – 1978 (R2004) Standard. ASME/ANSI B18.3.5M – 1986 (R2002) Standard. Reprinted from the standard listed by permission of the American Society of Mechanical Engineers. All rights reserved.

D.2.2) Hole Basis Transition and Interference Fits

Preferred Hole Basis Clearance Fits. Dimensions in mm.

Basic Size	Locational Transition		Locational Transition		Locational Interference		Medium Drive		Force	
	Hole H7	Shaft k6	Hole H7	Shaft n6	Hole H7	Shaft p6	Hole H7	Shaft s6	Hole H7	Shaft u6
1 max	1.010	1.006	1.010	1.010	1.010	1.012	1.010	1.020	1.010	1.024
min	1.000	1.000	1.000	1.004	1.000	1.006	1.000	1.014	1.000	1.018
1.2 max	1.210	1.206	1.210	1.210	1.210	1.212	1.210	1.220	1.210	1.224
min	1.200	1.200	1.200	1.204	1.200	1.206	1.200	1.214	1.200	1.218
1.6 max	1.610	1.606	1.610	1.610	1.610	1.612	1.610	1.620	1.610	1.624
min	1.600	1.600	1.600	1.604	1.600	1.606	1.600	1.614	1.600	1.618
2 max	2.010	2.006	2.010	2.020	2.010	2.012	2.010	2.020	2.010	2.024
min	2.000	2.000	2.000	2.004	2.000	2.006	2.000	1.014	2.000	2.018
2.5 max	2.510	2.506	2.510	2.510	2.510	2.512	2.510	2.520	2.510	2.524
min	2.500	2.500	2.500	2.504	2.500	2.506	2.500	2.514	2.500	2.518
3 max	3.010	3.006	3.010	3.010	3.010	3.012	3.010	3.020	3.010	3.024
min	3.000	3.000	3.000	3.004	3.000	3.006	3.000	3.014	3.000	3.018
4 max	4.012	4.009	4.012	4.016	4.012	4.020	4.012	4.027	4.012	4.031
min	4.000	4.001	4.000	4.008	4.000	4.012	4.000	4.019	4.000	4.023
5 max	5.012	5.009	5.012	5.016	5.012	5.020	5.012	5.027	5.012	5.031
min	5.000	5.001	5.000	5.008	5.000	5.012	5.000	5.019	5.000	5.023
6 max	6.012	6.009	6.012	6.016	6.012	6.020	6.012	6.027	6.012	6.031
min	6.000	6.001	6.000	6.008	6.000	6.012	6.000	6.019	6.000	6.023
8 max	8.015	8.010	8.015	8.019	8.015	8.024	8.015	8.032	8.015	8.037
min	8.000	8.001	8.000	8.010	8.000	8.015	8.000	8.023	8.000	8.028
10 max	10.015	10.010	10.015	10.019	10.015	10.024	10.015	10.032	10.015	10.037
min	10.000	10.001	10.000	10.010	10.000	10.015	10.000	10.023	10.000	10.028
12 max	12.018	12.012	12.018	12.023	12.018	12.029	12.018	12.039	12.018	12.044
min	12.000	12.001	12.000	12.012	12.000	12.018	12.000	12.028	12.000	12.033
16 max	16.018	16.012	16.018	16.023	16.018	16.029	16.018	16.039	16.018	16.044
min	16.000	16.001	16.000	16.012	16.000	16.018	16.000	16.028	16.000	16.033
20 max	20.021	20.015	20.021	20.028	20.021	20.035	20.021	20.048	20.021	20.054
min	20.000	20.002	20.000	20.015	20.000	20.022	20.000	20.035	20.000	20.041
25 max	25.021	25.015	25.021	25.028	25.021	25.035	25.021	25.048	25.021	25.061
min	25.000	25.002	25.000	25.015	25.000	25.022	25.000	25.035	25.000	25.048
30 max	30.021	30.015	30.021	30.028	30.021	30.035	30.021	30.048	30.021	30.061
min	30.000	30.002	30.000	30.015	30.000	30.022	30.000	30.035	30.000	30.048

ANSI B4.2 – 1978 (R2004) Standard. ASME/ANSI B18.3.5M – 1986 (R2002) Standard. Reprinted from the standard listed by permission of the American Society of Mechanical Engineers. All rights reserved.

D.2.3) Shaft Basis Clearance Fits

Preferred Shaft Basis Clearance Fits. Dimensions in mm.

Basic Size	Loose Running		Free Running		Close Running		Sliding		Locational Clearance	
	Hole H11	Shaft c11	Hole H9	Shaft d9	Hole H8	Shaft f7	Hole H7	Shaft g6	Hole H7	Shaft h6
1 max	1.120	1.000	1.045	1.000	1.020	1.000	1.012	1.000	1.010	1.000
min	1.060	0.940	1.020	0.975	1.006	0.990	1.002	0.994	1.000	0.994
1.2 max	1.320	1.200	1.245	1.200	1.220	1.200	1.212	1.200	1.210	1.200
min	1.260	1.140	1.220	1.175	1.206	1.190	1.202	1.194	1.200	1.194
1.6 max	1.720	1.600	1.645	1.600	1.620	1.600	1.612	1.600	1.610	1.600
min	1.660	1.540	1.620	1.575	1.606	1.590	1.602	1.594	1.600	1.594
2 max	2.120	2.000	2.045	2.000	2.020	2.000	2.012	2.000	2.010	2.000
min	2.060	1.940	2.020	1.975	2.006	1.990	2.002	1.994	2.000	1.994
2.5 max	2.620	2.500	2.545	2.500	2.520	2.500	2.512	2.500	2.510	2.500
min	2.560	2.440	2.520	2.475	2.506	2.490	2.502	2.494	2.500	2.494
3 max	3.120	3.000	3.045	3.000	3.020	3.000	3.012	3.000	3.010	3.000
min	3.060	2.940	3.020	2.975	3.006	2.990	3.002	2.994	3.000	2.994
4 max	4.145	4.000	4.060	4.000	4.028	4.000	4.016	4.000	4.012	4.000
min	4.070	3.925	4.030	3.970	4.010	3.988	4.004	3.992	4.000	3.992
5 max	5.145	5.000	5.060	5.000	5.028	5.000	5.016	5.000	5.012	5.000
min	5.070	4.925	5.030	4.970	5.010	4.988	5.004	4.992	5.000	4.992
6 max	6.145	6.000	6.060	6.000	6.028	6.000	6.016	6.000	6.012	6.000
min	6.070	5.925	6.030	5.970	6.010	5.988	6.004	5.992	6.000	5.992
8 max	8.170	8.000	8.076	8.000	8.035	8.000	8.020	8.000	8.015	8.000
min	8.080	7.910	8.040	7.964	8.013	7.985	8.005	7.991	8.000	7.991
10 max	10.170	10.000	10.076	10.000	10.035	10.000	10.020	10.000	10.015	10.000
min	10.080	9.910	10.040	9.964	10.013	9.985	10.005	9.991	10.000	9.991
12 max	12.205	12.000	12.093	12.000	12.043	12.000	12.024	12.000	12.018	12.000
min	12.095	11.890	12.050	11.957	12.016	11.982	12.006	11.989	12.000	11.989
16 max	16.205	16.000	16.093	16.000	16.043	16.000	16.024	16.000	16.018	16.000
min	16.095	15.890	16.050	15.957	16.016	15.982	16.006	15.989	16.000	15.989
20 max	20.240	20.000	20.117	20.000	20.053	20.000	20.028	20.000	20.021	20.000
min	20.110	19.870	20.065	19.948	20.020	19.979	20.007	19.987	20.000	19.987
25 max	25.240	25.000	25.117	25.000	25.053	25.000	25.028	25.000	25.021	25.000
min	25.110	24.870	25.065	24.948	25.020	24.979	25.007	24.987	25.000	24.987
30 max	30.240	30.000	30.117	30.000	30.053	30.000	30.028	30.000	30.021	30.000
min	30.110	29.870	30.065	29.948	30.020	29.979	30.007	29.987	30.000	29.987

ANSI B4.2 – 1978 (R2004) Standard. ASME/ANSI B18.3.5M – 1986 (R2002) Standard. Reprinted from the standard listed by permission of the American Society of Mechanical Engineers. All rights reserved.

D.2.4) <u>Shaft Basis Transition and Interference Fits</u>

Preferred Shaft Basis Transition and Interference Fits. Dimensions in mm.

Basic Size	Locational Transition		Locational Transition		Locational Interference		Medium Drive		Force	
	Hole H7	Shaft k6	Hole H7	Shaft n6	Hole H7	Shaft p6	Hole H7	Shaft s6	Hole H7	Shaft u6
1 max	1.000	1.000	0.996	1.000	0.994	1.000	0.986	1.000	0.982	1.000
min	0.990	0.994	0.986	0.994	0.984	0.994	0.976	0.994	0.972	0.994
1.2 max	1.200	1.200	1.196	1.200	1.194	1.200	1.186	1.200	1.182	1.200
min	1.190	1.194	1.186	1.194	1.184	1.194	1.176	1.194	1.172	1.194
1.6 max	1.600	1.600	1.596	1.600	1.594	1.600	1.586	1.600	1.582	1.600
min	1.590	1.594	1.586	1.594	1.584	1.594	1.576	1.594	1.572	1.594
2 max	2.000	2.000	1.996	2.000	1.994	2.000	1.986	2.000	1.982	2.000
min	1.990	1.994	1.986	1.994	1.984	1.994	1.976	1.994	1.972	1.994
2.5 max	2.500	2.500	2.496	2.500	2.494	2.500	2.486	2.500	2.482	2.500
min	2.490	2.494	2.486	2.494	2.484	2.494	2.476	2.494	2.472	2.494
3 max	3.000	3.000	2.996	3.000	2.994	3.000	2.986	3.000	2.982	3.000
min	2.990	2.994	2.986	2.994	2.984	2.994	2.976	2.994	2.972	2.994
4 max	4.003	4.000	3.996	4.000	3.992	4.000	3.985	4.000	3.981	4.000
min	3.991	5.992	3.984	5.992	3.980	5.992	3.973	5.992	3.969	5.992
5 max	5.003	5.000	4.996	5.000	4.992	5.000	4.985	5.000	4.981	5.000
min	4.991	4.992	4.984	4.992	4.980	4.992	4.973	4.992	4.969	4.992
6 max	6.003	6.000	5.996	6.000	5.992	6.000	5.985	6.000	5.981	6.000
min	5.991	5.992	5.984	5.992	5.980	5.992	5.973	5.992	5.969	5.992
8 max	8.005	8.000	7.996	8.000	7.991	8.000	7.983	8.000	7.978	8.000
min	7.990	7.991	7.981	7.991	7.976	7.991	7.968	7.991	7.963	7.991
10 max	10.005	10.000	9.996	10.000	9.991	10.000	9.983	10.000	9.978	10.000
min	9.990	9.991	9.981	9.991	9.976	9.991	9.968	9.991	9.963	9.991
12 max	12.006	12.000	11.995	12.000	11.989	12.000	11.979	12.000	11.974	12.000
min	11.988	11.989	11.977	11.989	11.971	11.989	11.961	11.989	11.956	11.989
16 max	16.006	16.000	15.995	16.000	15.989	16.000	15.979	16.000	15.974	16.000
min	15.988	15.989	15.977	15.989	15.971	15.989	15.961	15.989	15.956	15.989
20 max	20.006	20.000	19.993	20.000	19.986	20.000	19.973	20.000	19.967	20.000
min	19.985	19.987	19.972	19.987	19.965	19.987	19.952	19.987	19.946	19.987
25 max	25.006	25.000	24.993	25.000	24.986	25.000	24.973	25.000	24.960	25.000
min	24.985	24.987	24.972	24.987	24.965	24.987	24.952	24.987	24.939	24.987
30 max	30.006	30.000	29.993	30.000	29.986	30.000	29.973	30.000	29.960	30.000
min	29.985	29.987	29.972	29.987	29.965	29.987	29.952	29.987	29.939	29.987

ANSI B4.2 – 1978 (R2004) Standard. ASME/ANSI B18.3.5M – 1986 (R2002) Standard. Reprinted from the standard listed by permission of the American Society of Mechanical Engineers. All rights reserved.

NOTES:

APPENDIX E: THREADS AND FASTENERS

E.1) UNIFIED NATIONAL THREAD FORM

(External Threads) Approximate Minor diameter = $D - 1.0825P$ $\qquad$ P = Pitch

Nominal Size, in.	Basic Major Diameter (D)	Coarse UNC		Fine UNF		Extra Fine UNEF	
		Thds. Per in.	Tap Drill Dia.	Thds. Per in.	Tap Drill Dia.	Thds. Per in.	Tap Drill Dia.
#0	0.060	...	...	80	3/64	...	...
#1	0.0730	64	0.0595	72	0.0595	...	...
#2	0.0860	56	0.0700	64	0.0700	...	...
#3	0.0990	48	0.0785	56	0.0820	...	...
#4	0.1120	40	0.0890	48	0.0935	...	...
#5	0.1250	40	0.1015	44	0.1040	...	...
#6	0.1380	32	0.1065	40	0.1130	...	...
#8	0.1640	32	0.1360	36	0.1360	...	...
#10	0.1900	24	0.1495	32	0.1590	...	...
#12	0.2160	24	0.1770	28	0.1820	32	0.1850
1/4	0.2500	20	0.2010	28	0.2130	32	7/32
5/16	0.3125	18	0.257	24	0.272	32	9/32
3/8	0.3750	16	5/16	24	0.332	32	11/32
7/16	0.4375	14	0.368	20	25/64	28	13/32
1/2	0.5000	13	27/64	20	29/64	28	15/32
9/16	0.5625	12	31/64	18	33/64	24	33/64
5/8	0.6250	11	17/32	18	37/64	24	37/64
11/16	0.675	...	...	...	...	24	41/64
3/4	0.7500	10	21/32	16	11/16	20	45/64
13/16	0.8125	...	...	...	...	20	49/64
7/8	0.8750	9	49/64	14	13/16	20	53/64
15/16	0.9375	...	...	...	...	20	57/64
1	1.0000	8	7/8	12	59/64	20	61/64
1 1/8	1.1250	7	63/64	12	1 3/64	18	1 5/64
1 1/4	1.2500	7	1 7/64	12	1 11/64	18	1 3/16
1 3/8	1.3750	6	1 7/32	12	1 19/64	18	1 5/16
1 1/2	1.5000	6	1 11/32	12	1 27/64	18	1 7/16
1 5/8	1.6250	...	...	...	...	18	1 9/16
1 3/4	1.7500	5	1 9/16	...	...	...	...
1 7/8	1.8750	...	...	...	...	...	...
2	2.0000	4 1/2	1 25/32	...	...	...	...
2 1/4	2.2500	4 1/2	2 1/32	...	...	...	...
2 1/2	2.5000	4	2 1/4	...	...	...	...
2 3/4	2.7500	4	2 1/2	...	...	...	...

ASME B1.1 – 2003 Standard. Reprinted from the standard listed by permission of the American Society of Mechanical Engineers. All rights reserved.

E.2) **METRIC THREAD FORM**

(External Threads) Approximate Minor diameter = $D - 1.2075P$ P = Pitch

Preferred sizes for commercial threads and fasteners are shown in boldface type.

Coarse (general purpose)		Fine	
Nominal Size (D) & Thread Pitch	Tap Drill Diameter, mm	Nominal Size & Thread Pitch	Tap Drill Diameter, mm
M1.6 x 0.35	1.25	---	---
M1.8 x 0.35	1.45	---	---
M2 x 0.4	1.6	---	---
M2.2 x 0.45	1.75	---	---
M2.5 x 0.45	2.05	---	---
M3 x 0.5	2.5	---	---
M3.5 x 0.6	2.9	---	---
M4 x 0.7	3.3	---	---
M4.5 x 0.75	3.75	---	---
M5 x 0.8	4.2	---	---
M6 x 1	5.0	---	---
M7 x 1	6.0	**---**	**---**
M8 x 1.25	6.8	**M8 x 1**	7.0
M9 x 1.25	7.75	**---**	---
M10 x 1.5	8.5	**M10 x 1.25**	8.75
M11 x 1.5	9.50	---	---
M12 x 1.75	10.30	**M12 x 1.25**	10.5
M14 x 2	12.00	**M14 x 1.5**	12.5
M16 x 2	14.00	**M16 x 1.5**	14.5
M18 x 2.5	15.50	**M18 x 1.5**	16.5
M20 x 2.5	17.5	**M20 x 1.5**	18.5
M22 x 2.5[b]	19.5	**M22 x 1.5**	20.5
M24 x 3	21.0	**M24 x 2**	22.0
M27 x 3[b]	24.0	**M27 x 2**	25.0
M30 x 3.5	26.5	**M30 x 2**	28.0
M33 x 3.5	29.5	M33 x 2	31.0
M36 x 4	32.0	**M36 x 2**	33.0
M39 x 4	35.0	M39 x 2	36.0
M42 x 4.5	37.5	**M42 x 2**	39.0
M45 x 4.5	40.5	M45 x 1.5	42.0
M48 x 5	43.0	**M48 x 2**	45.0
M52 x 5	47.0	M52 x 2	49.0
M56 x 5.5	50.5	**M56 x 2**	52.0
M60 x 5.5	54.5	M60 x 1.5	56.0
M64 x 6	58.0	**M64 x 2**	60.0
M68 x 6	62.0	M68 x 2	64.0
M72 x 6	66.0	**M72 x 2**	68.0
M80 x 6	74.0	**M80 x 2**	76.0
M90 x 6	84.0	**M90 x 2**	86.0
M100 x 6	94.0	**M100 x 2**	96.0

[b]Only for high strength structural steel fasteners

ASME B1.13M – 2001 Standard. Reprinted from the standard listed by permission of the American Society of Mechanical Engineers. All rights reserved.

E.3) <u>FASTENERS (INCH SERIES)</u>

CAUTION! All fastener dimensions have a tolerance. Therefore, each dimension has a maximum and minimum value. Only one size for each dimension is given in this appendix. That is all that is necessary to complete the problems given in the "Threads and Fasteners" chapter. For both values, please refer to the standards noted.

E.3.1) <u>Dimensions of Hex Bolts and Heavy Hex Bolts</u>

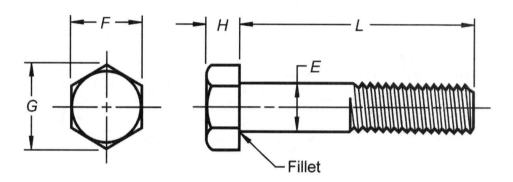

<u>Regular Hex Head Bolts</u>

Size (D)	Head Height Basic*	Width Across Flats Basic Adjust to sixteenths	Width Across Corners Max.
1/4	$H = 0.625\,D + 0.016$	$F = 1.500\,D + 0.062$	
5/16 – 7/16	$H = 0.625\,D + 0.016$		
1/2 – 7/8	$H = 0.625\,D + 0.031$		
1 – 1 7/8	$H = 0.625\,D + 0.062$	$F = 1.500\,D$	Max. $G = 1.1547\,F$
2 – 3 3/4	$H = 0.625\,D + 0.125$		
4	$H = 0.625\,D + 0.188$		

<u>Heavy Hex Head Bolts</u>

Size (D)	Head Height Basic*	Width Across Flats Basic Adjust to sixteenths	Width Across Corners Max.
1/2 - 3	Same as for regular hex head bolts.	$F = 1.500\,D + 0.125$	Max. $G = 1.1547\,F$

* Size to 1 in. adjusted to sixty-fourths. 1 1/8 through 2 1/2 in. sizes adjusted upward to thirty-seconds. 2 3/4 thru 4 in. sizes adjusted upward to sixteenths.

ASME B18.2.1 – 1996 Standard. Reprinted from the standard listed by permission of the American Society of Mechanical Engineers. All rights reserved.

E.3.2) <u>Dimensions of Hex Nuts and Hex Jam Nuts</u>

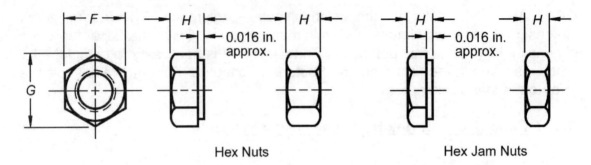

Hex Nuts   Hex Jam Nuts

<u>Hex Nuts</u>

Nut Size (D)	Nut Thickness Basic*	Width Across Flats Basic Adjust to sixteenths	Width Across Corners Max.
1/4	$H = 0.875 D$	$F = 1.500 D + 0.062$	
5/16 – 5/8	$H = 0.875 D$		Max. $G = 1.1547 F$
3/4 – 1 1/8	$H = 0.875 D - 0.016$	$F = 1.500 D$	
1 1/4 – 1 1/2	$H = 0.875 D - 0.031$		

<u>Hex Thick Nuts</u>

Nut Size (D)	Width Across Flats Basic Adjust to sixteenths	Width Across Corners Max.	Nut Thickness Basic
1/4	$F = 1.500 D + 0.062$		
5/16 – 5/8	$F = 1.500 D$	Max. $G = 1.1547 F$	See Table
3/4 – 1 1/2	$F = 1.500 D$		

Nut Size (D)	1/4	5/16	3/8	7/16	1/2	9/16	5/8
Nut Thickness Basic	9/32	21/64	13/32	29/64	9/16	39/64	23/32

Nut Size (D)	3/4	7/8	1	1 1/8	1 1/4	1 3/8	1 1/2
Nut Thickness Basic	13/16	29/32	1	1 5/32	1 1/4	1 3/8	1 1/2

ASME/ANSI B18.2.2 – 1987 (R1999) Standard. Reprinted from the standard listed by permission of the American Society of Mechanical Engineers. All rights reserved.

Hex Jam Nut

Nut Size (D)	Nut Thickness Basic*	Width Across Flats Basic Adjust to sixteenths	Width Across Corners Max.
1/4	See Table	$F = 1.500\ D + 0.062$	
5/16 – 5/8	See Table		Max. $G = 1.1547\ F$
3/4 – 1 1/8	$H = 0.500\ D - 0.047$	$F = 1.500\ D$	
1 1/4 – 1 1/2	$H = 0.500\ D - 0.094$		

Nut Size (D)	1/4	5/16	3/8	7/16	1/2	9/16	5/8
Nut Thickness Basic	5/32	3/16	7/32	1/4	5/16	5/16	3/8

ASME/ANSI B18.2.2 – 1987 (R1999) Standard. Reprinted from the standard listed by permission of the American Society of Mechanical Engineers. All rights reserved.

E.3.3) Dimensions of Hexagon and Spline Socket Head Cap Screws

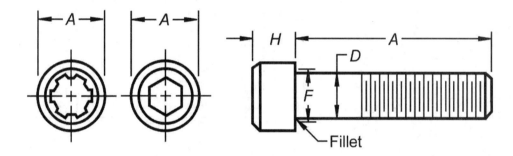

Screw Size (D)	Head Diameter	Head Height
#0 - #10	See Table	Max. $H = D$
1/4 - 4	Max. $A = 1.50\ D$	

Screw Size (D)	#0	#1	#2	#3	#4
Max. Head Diameter (A)	0.096	0.118	0.140	0.161	0.183

Screw Size (D)	#5	#6	#8	#10
Max. Head Diameter (A)	0.205	0.226	0.270	0.312

ASME B18.3 – 2003 Standard. Reprinted from the standard listed by permission of the American Society of Mechanical Engineers. All rights reserved.

E.3.4) Drill and Counterbore Sizes for Socket Head Cap Screws

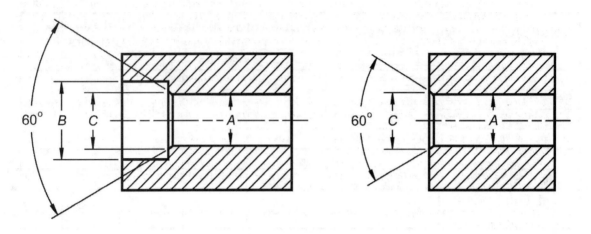

Nominal Size	Nominal Drill Size (A)		Counterbore	Countersink
of Screw (D)	Close Fit	Normal Fit	Diameter (B)	(C)
#0 (0.0600)	(#51) 0.067	(#49) 0.073	1/8	0.074
#1 (0.0730)	(#46) 0.081	(#43) 0.089	5/32	0.087
#2 (0.0860)	3/32	(#36) 0.106	3/16	0.102
#3 (0.0990)	(#36) 0.106	(#31) 0.120	7/32	0.115
#4 (0.1120)	1/8	(#29) 0.136	7/32	0.130
#5 (0.1250)	9/64	(#23) 0.154	1/4	0.145
#6 (0.1380)	(#23) 0.154	(#18) 0.170	9/32	0.158
#8 (0.1640)	(#15) 0.180	(#10) 0.194	5/16	0.188
#10 (0.1900)	(#5) 0.206	(#2) 0.221	3/8	0.218
1/4	17/64	9/32	7/16	0.278
5/16	21/64	11/32	17/32	0.346
3/8	25/64	13/32	5/8	0.415
7/16	29/64	15/32	23/32	0.483
1/2	33/64	17/32	13/16	0.552
5/8	41/64	21/32	1	0.689
3/4	49/64	25/32	1 3/16	0.828
7/8	57/64	29/32	1 3/8	0.963
1	1 1/64	1 1/32	1 5/8	1.100
1 1/4	1 9/32	1 5/16	2	1.370
1 1/2	1 17/32	1 9/16	2 3/8	1.640
1 3/4	1 25/32	1 13/16	2 3/4	1.910
2	2 1/32	2 1/16	3 1/8	2.180

Notes on next page.

Notes:
(1) *Countersink.* It is considered good practice to countersink or break the edges of holes that are smaller than F (max.) in parts having a hardness which approaches, equals, or exceeds the screw hardness. If such holes are not countersunk, the heads of screws may not seat properly or the sharp edges on hols may deform the fillets on screws thereby making them susceptible to fatigue in applications involving dynamic loading. The countersink or corner relief, however, should not be larger than is necessary to insure that the fillet on the screw is cleared. Normally, the diameter of countersink does not heave to exceed F (max.). Countersinks or corner reliefs in excess of this diameter reduce the effective bearing area and introduce the possibility of imbedment where the parts to be fastened are softer than the screws or brinnelling or flaring of the heads of the screws where the parts to be fastened are harder than the screws.
(2) *Close Fit.* The close fit is normally limited to holes for those lengths of screws that are threaded to the head in assemblies where only one screw is to be used or where two or more screws are to be used and the mating holes are to be produced either at assembly or by matched and coordinated tooling.
(3) *Normal Fit.* The normal fit is intended for screws of relatively long length or for assemblies involving two or more screws where the mating holes are to be produced by conventional tolerancing methods. It provides for the maximum allowable eccentricity of the longest standard screws and for certain variations in the parts to be fastened, such as: deviations in hole straightness, angularity between the axis of the tapped hole and that of the hole for the shank, differences in center distances of the mating holes, etc.

ASME B18.3 – 2003 Standard. Reprinted from the standard listed by permission of the American Society of Mechanical Engineers. All rights reserved.

E.3.5) Dimensions of Hexagon and Spline Socket Flat Countersunk Head Cap Screws

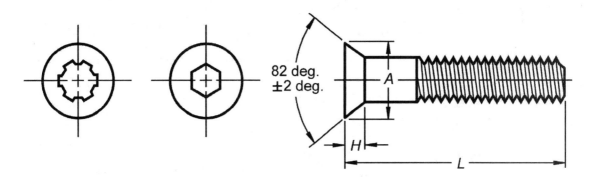

Screw Size (D)	Head Diameter (A) Theor. Sharp	Head Height (H)
#0 - #3	See Table	Max. $H = 0.5$ (Max. $A - D$) * cot (41°)
#4 – 3/8	Max. $A = 2D + 0.031$	
7/16	Max. $A = 2D - 0.031$	
1/2 – 1 1/2	Max. $A = 2D - 0.062$	

ASME B18.3 – 2003 Standard. Reprinted from the standard listed by permission of the American Society of Mechanical Engineers. All rights reserved.

E.3.6) <u>Dimensions of Slotted Flat Countersunk Head Cap Screws</u>

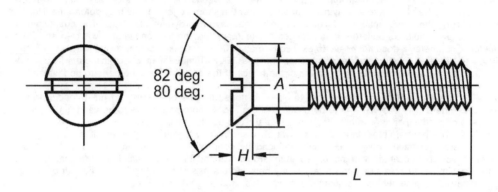

Screw Size (D)	Head Diameter (A) Thero. Sharp	Head Height (H)
1/4 through 3/8	Max. $A = 2.000\ D$	Max. $H = 0.596\ D$
7/16	Max. $A = 2.000\ D - 0.063$	Max. $H = 0.596\ D - 0.0375$
1/2 through 1	Max. $A = 2.000\ D - 0.125$	Max. $H = 0.596\ D - 0.075$
1 1/8 through 1 1/2	Max. $A = 2.000\ D - 0.188$	Max. $H = 0.596\ D - 0.112$

ASME B18.6.2 – 1998 Standard. Reprinted from the standard listed by permission of the American Society of Mechanical Engineers. All rights reserved.

E.3.7) <u>Dimensions of Slotted Round Head Cap Screws</u>

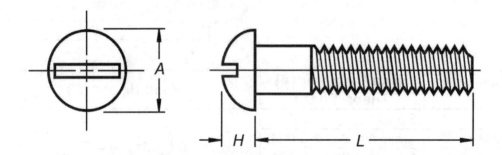

Screw Size (D)	Head Diameter (A) Thero. Sharp	Head Height (H)
1/4 and 5/16	Max. $A = 2.000\ D - 0.063$	Max. $H = 0.875\ D - 0.028$
3/8 and 7/16	Max. $A = 2.000\ D - 0.125$	Max. $H = 0.875\ D - 0.055$
1/2 and 9/16	Max. $A = 2.000\ D - 0.1875$	Max. $H = 0.875\ D - 0.083$
5/8 and 3/4	Max. $A = 2.000\ D - 0.250$	Max. $H = 0.875\ D - 0.110$

ASME B18.6.2 – 1998 Standard. Reprinted from the standard listed by permission of the American Society of Mechanical Engineers. All rights reserved.

E.3.8) <u>Dimensions of Preferred Sizes of Type A Plain Washers</u>

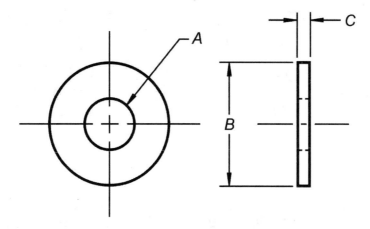

Nominal Washer Size[*]	Inside Diameter (A) Basic	Outside Diameter (B) Basic	Thickness (C)	Nominal Washer Size[*]	Inside Diameter (A) Basic	Outside Diameter (B) Basic	Thickness (C)
	0.078	0.188	0.020	1 N	1.062	2.000	0.134
	0.094	0.250	0.020	1 W	1.062	2.500	0.165
	0.125	0.312	0.032	1 1/8 N	1.250	2.250	0.134
#6 (0.138)	0.156	0.375	0.049	1 1/8 W	1.250	2.750	0.165
#8 (0.164)	0.188	0.438	0.049	1 1/4 N	1.375	2.500	0.165
#10 (0.190)	0.219	0.500	0.049	1 1/4 W	1.375	3.000	0.165
3/16	0.250	0.562	0.049	1 3/8 N	1.500	2.750	0.165
#12 (0.216)	0.250	0.562	0.065	1 3/8 W	1.500	3.250	0.180
1/4 N	0.281	0.625	0.065	1 1/2 N	1.625	3.000	0.165
1/4 W	0.312	0.734	0.065	1 1/2 W	1.625	3.500	0.180
5/16 N	0.344	0.688	0.065	1 5/8	1.750	3.750	0.180
5/16 W	0.375	0.875	0.083	1 3/4	1.875	4.000	0.180
3/8 N	0.406	0.812	0.065	1 7/8	2.000	4.250	0.180
3/8 W	0.438	1.000	0.083	2	2.125	4.500	0.180
7/16 N	0.469	0.922	0.065	2 1/4	2.375	4.750	0.220
7/16 W	0.500	1.250	0.083	2 1/2	2.625	5.000	0.238
1/2 N	0.531	1.062	0.095	2 3/4	2.875	5.250	0.259
1/2 W	0.562	1.375	0.109	3	3.125	5.500	0.284
9/16 N	0.594	1.156	0.095				
9/16 W	0.625	1.469	0.109				
5/8 N	0.656	1.312	0.095				
5/8 W	0.688	1.750	0.134				
3/4 N	0.812	1.469	0.134				
3/4 W	0.812	2.000	0.148				
7/8 N	0.938	1.750	0.134				
7/8 W	0.938	2.250	0.165				

[*] Nominal washer sizes are intended for use with comparable nominal screw or bolt sizes.

ANSI B18.22.1 – 1965 (R2003) Standard. Reprinted from the standard listed by permission of the American Society of Mechanical Engineers. All rights reserved.

E.3.9) **Dimensions of Regular Helical Spring-Lock Washers**

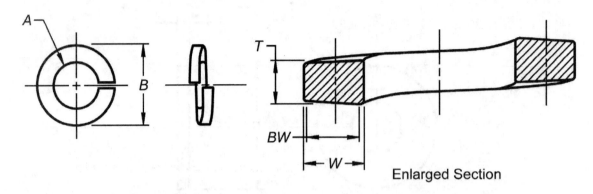

Enlarged Section

Nominal Washer Size	Min. Inside Diameter (A)	Max. Outside Diameter (B)	Mean Section Thickness (T)	Min. Section Width (W)	Min. Bearing Width (BW)
#2 (0.086)	0.088	0.172	0.020	0.035	0.024
#3 (0.099)	0.101	0.195	0.025	0.040	0.028
#4 (0.112)	0.114	0.209	0.025	0.040	0.028
#5 (0.125)	0.127	0.236	0.031	0.047	0.033
#6 (.0138)	0.141	0.250	0.031	0.047	0.033
#8 (0.164)	0.167	0.293	0.040	0.055	0.038
#10 (0.190)	0.193	0.334	0.047	0.062	0.043
#12 (0.216)	0.220	0.377	0.056	0.070	0.049
1/4	0.252	0.487	0.062	0.109	0.076
5/16	0.314	0.583	0.078	0.125	0.087
3/8	0.377	0.680	0.094	0.141	0.099
7/16	0.440	0.776	0.109	0.156	0.109
1/2	0.502	0.869	0.125	0.171	0.120
9/16	0.564	0.965	0.141	0.188	0.132
5/8	0.628	1.073	0.156	0.203	0.142
11/16	0.691	1.170	0.172	0.219	0.153
3/4	0.753	1.265	0.188	0.234	0.164
13/16	0.816	1.363	0.203	0.250	0.175
7/8	0.787	1.459	0.219	0.266	0.186
15/16	0.941	1.556	0.234	0.281	0.197
1	1.003	1.656	0.250	0.297	0.208
1 1/16	1.066	1.751	0.266	0.312	0.218
1 1/8	1.129	1.847	0.281	0.328	0.230
1 3/16	1.192	1.943	0.297	0.344	0.241
1 1/4	1.254	2.036	0.312	0.359	0.251
1 5/16	1.317	2.133	0.328	0.375	0.262
1 3/8	1.379	2.219	0.344	0.391	0.274
1 7/16	1.442	2.324	0.359	0.406	0.284
1 1/2	1.504	2.419	0.375	0.422	0.295
1 5/8	1.633	2.553	0.389	0.424	0.297
1 3/4	1.758	2.679	0.389	0.424	0.297
1 7/8	1.883	2.811	0.422	0.427	0.299
2	2.008	2.936	0.422	0.427	0.299
2 1/4	2.262	3.221	0.440	0.442	0.309
2 1/2	2.512	3.471	0.440	0.422	0.309
2 3/4	2.762	3.824	0.458	0.491	0.344
3	3.012	4.074	0.458	0.491	0.344

ASME B18.21.1 – 1999 Standard. Reprinted from the standard listed by permission of the American Society of Mechanical Engineers. All rights reserved.

E.4) __METRIC FASTENERS__

E.4.1) __Dimensions of Hex Bolts__

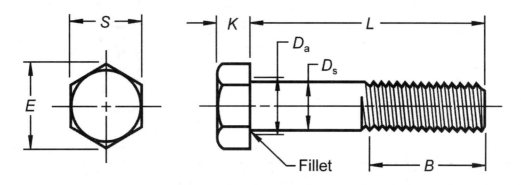

D	D_s	S	E	K	D_a	Thread Length (B)		
Nominal Bolt Diameter and Thread Pitch	Max. Body Diameter	Max. Width Across Flats	Max. Width Across Corners	Max. Head Height	Fillet Transition Diameter	Bolt Lengths ≤ 125	Bolt Lengths > 125 and ≤ 200	Bolt Lengths > 200
M5 x 0.8	5.48	8.00	9.24	3.88	5.7	16	22	35
M6 x 1	6.19	10.00	11.55	4.38	6.8	18	24	37
M8 x 1.25	8.58	13.00	15.01	5.68	9.2	22	28	41
M10 x 1.5	10.58	16.00	18.48	6.85	11.2	26	32	45
M12 x 1.75	12.70	18.00	20.78	7.95	13.7	30	36	49
M14 x 2	14.70	21.00	24.25	9.25	15.7	34	40	53
M16 x 2	16.70	24.00	27.71	10.75	17.7	38	44	57
M20 x 2.5	20.84	30.00	34.64	13.40	22.4	46	52	65
M24 x 3	24.84	36.00	41.57	15.90	26.4	54	60	73
M30 x 3.5	30.84	46.00	53.12	19.75	33.4	66	72	85
M36 x 4	37.00	55.00	63.51	23.55	39.4	78	84	97
M42 x 4.5	43.00	65.00	75.06	27.05	45.4	90	96	109
M48 x 5	49.00	75.00	86.60	31.07	52.0	102	108	121
M56 x 5.5	57.00	85.00	98.15	36.20	62.0		124	137
M64 x 6	65.52	95.00	109.70	41.32	70.0		140	153
M72 x 6	73.84	105.00	121.24	46.45	78.0		156	169
M80 x 6	82.16	115.00	132.79	51.58	86.0		172	185
M90 x 6	92.48	130.00	150.11	57.74	96.0		192	205
M100 x 6	102.80	145.00	167.43	63.90	107.0		212	225

ASME/ANSI B18.3.5M – 1986 (R2002) Standard. Reprinted from the standard listed by permission of the American Society of Mechanical Engineers. All rights reserved.

E.4.2) Dimensions of Hex Nuts, Style 1

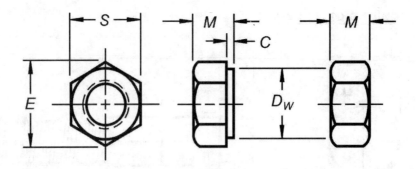

D	S	E	M	D_W	C
Nominal Bolt Diameter and Thread Pitch	Max. Width Across Flats	Max. Width Across Corners	Max. Thickness	Min. Bearing Face Diameter	Max. Washer Face Thickness
M1.6 x 0.35	3.20	3.70	1.30	2.3	
M2 x 0.4	4.00	4.62	1.60	3.1	
M2.5 x 0.45	5.00	5.77	2.00	4.1	
M3 x 0.5	5.50	6.35	2.40	4.6	
M3.5 x 0.6	6.00	6.93	2.80	5.1	
M4 x 0.7	7.00	8.08	3.20	6.0	
M5 x 0.8	8.00	9.24	4.70	7.0	
M6 x 1	10.00	11.55	5.20	8.9	
M8 x 1.25	13.00	15.01	6.80	11.6	
M10 x 1.5	15.00	17.32	9.10	13.6	
M10 x 1.5	16.00	18.45	8.40	14.6	
M12 x 1.75	18.00	20.78	10.80	16.6	
M14 x 2	21.00	24.25	12.80	19.4	
M16 x 2	24.00	27.71	14.80	22.4	
M20 x 2.5	30.00	34.64	18.00	27.9	0.8
M24 x 3	36.00	41.57	21.50	32.5	0.8
M30 x 3.5	46.00	53.12	25.60	42.5	0.8
M36 x 4	55.00	63.51	31.00	50.8	0.8

ASME B18.2.4.1M – 2002 Standard. Reprinted from the standard listed by permission of the American Society of Mechanical Engineers. All rights reserved.

E.4.3) <u>Dimensions of Metric Socket Head Cap Screws</u>

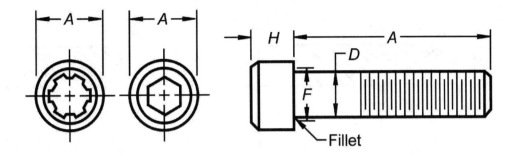

Dimensions in mm

Screw Size (*D*)	Head Diameter (*A*)	Head Height (*H*)
1.6 through 2.5	See Table	
3 through 8	Max. *A* = 1.5 *D* + 1	Max. *H* = *D*
> 10	Max. *A* = 1.5 *D*	

Screw Size (*D*)	1.6	2	2.5
Max. Head Diameter (*A*)	3.00	3.80	4.50

ASME/ANSI B18.3.1M – 1986 (R2002) Standard. Reprinted from the standard listed by permission of the American Society of Mechanical Engineers. All rights reserved.

E.4.4) Drill and Counterbore Sizes for Socket Head Cap Screws

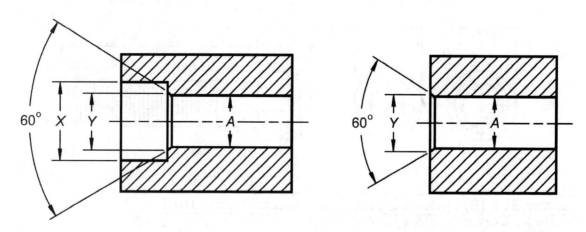

Nominal Size or Basic Screw Diameter	A		X	Y
	Nominal Drill Size		Counterbore Diameter	Countersink Diameter
	Close Fit	Normal Fit		
M1.6	1.80	1.95	3.50	2.0
M2	2.20	2.40	4.40	2.6
M2.5	2.70	3.00	5.40	3.1
M3	3.40	3.70	6.50	3.6
M4	4.40	4.80	8.25	4.7
M5	5.40	5.80	9.75	5.7
M6	6.40	6.80	11.25	6.8
M8	8.40	8.80	14.25	9.2
M10	10.50	10.80	17.25	11.2
M12	12.50	12.80	19.25	14.2
M14	14.50	14.75	22.25	16.2
M16	16.50	16.75	25.50	18.2
M20	20.50	20.75	31.50	22.4
M24	24.50	24.75	37.50	26.4
M30	30.75	31.75	47.50	33.4
M36	37.00	37.50	56.50	39.4
M42	43.00	44.00	66.00	45.6
M48	49.00	50.00	75.00	52.6

ASME/ANSI B18.3.1M – 1986 (R2002) Standard. Reprinted from the standard listed by permission of the American Society of Mechanical Engineers. All rights reserved.

E.4.5) Dimensions of Metric Countersunk Socket Head Cap Screws

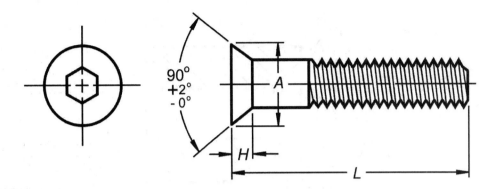

Basic Screw Diameter and Thread Pitch	Head Diameter (*A*) Theor. Sharp	Head Height (*H*)
M3 x 0.5	6.72	1.86
M4 x 0.7	8.96	2.48
M5 x 0.8	11.20	3.10
M6 x 1	13.44	3.72
M8 x 1.25	17.92	4.96
M10 x 1.5	22.40	6.20
M12 x 1.75	26.88	7.44
M14 x 2	30.24	8.12
M16 x 2	33.60	8.80
M20 x 2.5	40.32	10.16

ASME/ANSI B18.3.5M – 1986 (R2002) Standard. Reprinted from the standard listed by permission of the American Society of Mechanical Engineers. All rights reserved.

E.4.5) <u>Drill and Countersink Sizes for Flat Countersunk Head Cap Screws</u>

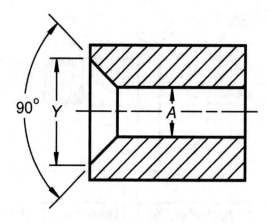

D	A	Y
Nominal Screw Size	Nominal Hole Diameter	Min. Countersink Diameter
M3	3.5	6.72
M4	4.6	8.96
M5	6.0	11.20
M6	7.0	13.44
M8	9.0	17.92
M10	11.5	22.40
M12	13.5	26.88
M14	16.0	30.24
M16	18.0	33.60
M20	22.4	40.32

ASME/ANSI B18.3.5M – 1986 (R2002) Standard. Reprinted from the standard listed by permission of the American Society of Mechanical Engineers. All rights reserved.

E.5) BOLT AND SCREW CLEARANCE HOLES

E.5.1) Inch Clearance Holes

Nominal Screw Size	Fit Classes		
	Normal	Close	Loose
	Nominal Drill Size		
#0	#48	#51	3/32
#1	#43	#46	#37
#2	#38	3/32	#32
#3	#32	#36	#30
#4	#30	#31	#27
#5	5/32	9/64	11/64
#6	#18	#23	#13
#8	#9	#15	#3
#10	#2	#5	B
1/4	9/32	17/64	19/64
5/16	11/32	21/64	23/64
3/8	13/32	25/64	27/64
7/16	15/32	29/64	31/64
1/2	9/16	17/32	39/64
5/8	11/16	21/32	47/64
3/4	13/16	25/32	29/32
7/8	15/16	29/32	1 1/32
1	1 3/32	1 1/32	1 5/32
1 1/8	1 7/32	1 5/32	1 5/16
1 1/4	1 11/32	1 9/32	1 7/16
1 3/8	1 1/2	1 7/16	1 39/64
1 1/2	1 5/8	1 9/16	1 47/64

ASME B18.2.8 – 1999 Standard. Reprinted from the standard listed by permission of the American Society of Mechanical Engineers. All rights reserved.

E.5.2) <u>Metric Clearance Holes</u>

Nominal Screw Size	Fit Classes		
	Normal	Close	Loose
	Nominal Drill Size		
M1.6	1.8	1.7	2
M2	2.4	2.2	2.6
M2.5	2.9	2.7	3.1
M3	3.4	3.2	3.6
M4	4.5	4.3	4.8
M5	5.5	5.3	5.8
M6	6.6	6.4	7
M8	9	8.4	10
M10	11	10.5	12
M12	13.5	13	14.5
M14	15.5	15	16.5
M16	17.5	17	18.5
M20	22	21	24
M24	26	25	28
M30	33	31	35
M36	39	37	42
M42	45	43	48
M48	52	50	56
M56	62	58	66
M64	70	66	74
M72	78	74	82
M80	86	82	91
M90	96	93	101
M100	107	104	112

ASME B18.2.8 – 1999 Standard. Reprinted from the standard listed by permission of the American Society of Mechanical Engineers. All rights reserved.

APPENDIX F: REFERENCES

- ASME B1.1 – 2003: Unified Inch Screw Threads (UN and UNR Thread Form)
- ASME B1.13M – 2001: Metric Screw Threads: M Profile
- USAS/ASME B4.1 – 1967 (R2004): Preferred Limits and Fits for Cylindrical Parts
- ANSI B4.2 – 1978 (R2004): Preferred Metric Limits and Fits
- ASME B18.2.1 – 1996: Square and Hex Bolts and Screws (Inch Series)
- ASME/ANSI B18.2.2 – 1987 (R1999): Square and Hex Nuts (Inch Series)
- ANSI B18.2.3.5M – 1979 (R2001): Metric Hex Bolts
- ASME B18.2.4.1M – 2002: Metric Hex Nuts, Style 1
- ASME B18.2.8 – 1999: Clearance Holes for Bolts, Screws, and Studs
- ASME B18.3 – 2003: Socket Cap, Shoulder, and Set Screws, Hex and Spline Keys (Inch Series)
- ASME/ANSI B18.3.1M – 1986 (R2002): Socket Head Cap Screws (Metric Series)
- ASME/ANSI B18.3.5M – 1986 (R2002): Hexagon Socket Flat Countersunk Head Cap Screws (Metric Series)
- ASME 18.6.2 – 1998: Slotted Head Cap Screws, Square Head Set Screws, and Slotted Headless Set Screws (Inch Series)
- ASME B18.21.1 – 1999: Lock Washers (Inch Series)
- ANSI B18.22.1 – 1965 (R2003): Plain Washers
- ASME Y14.2M – 1992 (R2003): Line Conventions and Lettering
- ASME Y14.3 – 2003: Multiview and Sectional view Drawings
- ASME Y14.5M – 1994: Dimensioning and Tolerancing
- ASME Y14.6 – 2001: Screw Thread Representation
- ASME Y14.100 – 2000: Engineering Drawing Practices
- 26th Edition of the Machinery's Handbook

NOTES:

NOTES:

NOTES:

NOTES:

NOTES:

NOTES:

NOTES:

NOTES: